HOW TO RAISE GOATS

EVERYTHING YOU NEED TO KNOW

MEAT, MILK, FIBER & PET GOATS
BREED GUIDE & PURCHASING
PROPER CARE & HEALTHY FEEDING
SHOWING ADVICE

Carol A. Amundson

Voyageur Press

First published in 2009 by Voyageur Press, an imprint of MBI Publishing Company, 400 First Avenue North, Suite 300, Minneapolis, MN 55401 USA

Voyageur Press titles are also available at discounts in bulk quantity for industrial or sales-promotional use. For details write to Special Sales Manager at MBI Publishing Company, 400 First Avenue North, Suite 300, Minneapolis, MN 55401 USA.

To find out more about our books, visit us online at www.voyageurpress.com.

Library of Congress Cataloging-in-Publication Data

Amundson, Carol A.
  How to raise goats / Carol A. Amundson.
    p. cm.
  Includes index.
  ISBN-13: 978-0-7603-3157-6 (softbound)
  1. Goats. I. Title.
  SF383.A68 2008
  636.3'9—dc22

                              2007043097

Edited by Danielle Ibister
Designed by Kazuko Collins
Diagrams by Lindsay Haas

Printed in Singapore

Front cover main photograph
    © *Norvia Behling*
Front cover inset photograph
    © *Jen Brown*
Back cover and spine photographs
    © *Shutterstock*

CONTENTS

ACKNOWLEDGMENTS 4

INTRODUCTION 6

CHAPTER 1:   A GOAT IS JUST A GOAT—RIGHT? 8

CHAPTER 2:   A PHYSICAL AND BEHAVIORAL OVERVIEW 24

CHAPTER 3:   BEFORE BRINGING YOUR GOATS HOME 34

CHAPTER 4:   GETTING YOUR GOATS 40

CHAPTER 5:   HOUSING YOUR GOATS 46

CHAPTER 6:   FEEDING YOUR GOATS 54

CHAPTER 7:   BREEDING YOUR GOATS 66

CHAPTER 8:   KIDS AND KIDDING 78

CHAPTER 9:   KEEPING HEALTHY GOATS 92

CHAPTER 10:   DEALING WITH SICK GOATS 106

CHAPTER 11:   TRAVELING WITH YOUR GOATS 124

CHAPTER 12:   KEEPING MILK GOATS 128

CHAPTER 13:   KEEPING MEAT GOATS 134

CHAPTER 14:   KEEPING FIBER GOATS 138

CHAPTER 15:   ENJOYING YOUR PET GOATS 144

CHAPTER 16:   SHOWING YOUR GOATS 150

CHAPTER 17:   MARKETING YOUR GOATS AND THEIR PRODUCTS 158

APPENDICES   I. MANAGEMENT CALENDAR 164

              II. ASSOCIATIONS, CLUBS, AND REGISTRIES 168

              III. EQUIPMENT AND SUPPLY DEALERS 180

              IV. PRINT RESOURCES 182

              V. INTERNET RESOURCES 183

              VI. GLOSSARY: HOW TO TALK GOAT 184

INDEX 190

ABOUT THE AUTHOR 192

ACKNOWLEDGMENTS

I am fortunate to have the support of many people. Everyone who reads this book will be getting not just my knowledge, but that of the people who helped me along the way.

Many of the photos between these pages were taken by me as well as family friends. Some have been gleaned from other breeders. I could not have completed this project without the wonderful photography of my friend Jennifer Brown. Jen came to the farm to get a kitten. Her camera work with the animals during her visit was so inspired that I pleaded with her to take photos for the book. She even stuck with me through some of the more disturbing medical procedures!

David Weber, my partner in the market garden venture Farm Fresh Veg, started Cutter Farms Goats in 2006. His herd and the Terrapin Acres goats comingle in another joint venture. Dave's help performing herd upkeep and chores has made things a lot easier around here. As I've answered Dave's questions and shown him goat management techniques, he's helped revise and add to this book. He's also proven to be an able photographer.

Poplar Hill will always seem like home after three years in a cooperative effort with Vince and Christine Maefksy. I have met quite a few goat owners who started with one or two pet goats and now face a hobby gone wild. One of these breeders is Vince Maefsky. Poplar Hill Dairy Goats—the largest goat dairy in Minnesota and one of the largest in the country—is only a few miles from our farm. I met the Maefsky family through the Minnesota Dairy Goat Association. (I also co-owned a buck with their daughter, Sarah Johnson.) Vincent picked up milk at our farm twice a week during my first commercial venture. I learned much of my dairy knowledge through Vince's stories and travels as he put up his beautiful dairy barn. Later, after our cheese plant no longer purchased milk from producers and we moved to another location, I took Vince and Chris up on their offer to move my milkers to Poplar Hill. Milking five mornings a week from three to seven (the time it takes to milk some five hundred goats in a forty-head modern milking parlor) was an education worth a degree. Vince and Chris, with their experience raising goats for more than thirty-five years, should be the ones writing this book—except I suspect they are too busy!

Photographer Jennifer Brown takes time out from behind her camera to visit with goats.
Barb O'Meehan

My friend Anna Boll Johnson was the first child I guided into the world of goats. Anna started helping on my farm at the age of thirteen. Someday I hope to be able to show her new daughter, Elsie, some descendants of the animals that her mother showed. My daughter, Viveka, twelve, now provides me with firsthand knowledge of animal husbandry as seen through the eyes of children. Kids—human and goat—go together naturally.

During my years raising and learning about goats, I have had the opportunity to get to know a lot of fascinating people. I can't even begin to thank those who gave me my first lessons in goat care. I am indebted to my friend Lynn Litterer, who took me to a class at the University of Minnesota about raising Angora goats—long before I moved to the country or had space for them. The Minnesota Dairy Goat Association, whose newsletter (edited at the time by Mark Boorsma) taught me some basics, led me to ads for breeders so I could buy my first animals, and then trapped me into the editor position for three and a half fascinating years.

I have fond memories of the breeders of my first animals. Maddie Frounfelter of Landmark Oaks taught me how to milk and sold me Celestial Star. Linda Libra provided my second animal, Marshland DD Menolly.

Linda became a friend and mentor and also allowed her daughters to help with my herd. Gloria Splinter of Hidden Springs Game Farm shared her exotic livestock expertise about Pygmies and Nigerians and introduced me to my first fainting goats when her husband, Roland, brought some home from a sale in Iowa. Gloria and her vet tech daughter, Tara, even listened and gave ideas as I read parts of this book aloud over the phone.

At goat meetings and shows, I met people who will always be available to me for advice and help. Lucresha Larson, "the midnight milker," has a full medicine chest and is willing to share—as long as I don't call much before noon. Bev and R. J. Nohr of Charis Manor and Deb and Sharla Macke of Raintree and Calico Acres Dairy Goats are breeders I can always call on when I need a shoulder to cry on or have a triumph to share.

I wish to extend thanks to Annette Maze of Hill Country Farms. She graciously allowed a drop-in visitor to have a complete tour of her ranch at the time of the first American Boer Goat Association Convention in Texas in 1994. Visiting a full-fledged embryo transplant facility and hearing her many stories about her years working with goats added another valuable layer of experience.

With the advent of the Internet, I have been able to expand my knowledge exponentially. With a click of a mouse, I can ask goat owners across the country for help with a sick animal. I highly recommend this resource, with the caveat to read a lot—and don't believe everything you find! See the appendix for a list of Internet resources.

Many thanks particularly to Linda Campbell of Khimaira Farms for her advice and help. Molly and Larry Bunton of Fiasco Farms designed my farm logo. I owe them even more for the incredible job they do on their website explaining basic goat care and sharing years of expertise. While I have never spoken to Joyce Lazarro of Saanendoah, her information on copper deficiency in goats helped me tremendously, and I highly recommend her resources.

My love and admiration for my husband, Wayne, and his willingness to be a part of this insane zoo called Terrapin Acres can't be counted! Almost as common as the fiercely passionate goat owner is the non-animal-loving spouse in the background. I have known marriages to dissolve and herds to disappear under the weight of 365-days-a-year animal care. Wayne has stuck with me. In spite of his lack of enthusiasm for the creatures themselves, Wayne helps with chores, shows goats (when necessary), and has even served several terms as Show Secretary. His broad shoulders catch my tears when an animal dies. Along the way, I have spent more money on goats than he wants to think about. Finally, when I thought Wayne had heard about as much "goat talk" as he could stand, he encouraged me to write this book! Without him, this book would not exist.

My husband, Wayne, pitches in with everything from hay feeding to milking.
Jen Brown

Goats are naturally friendly.

INTRODUCTION

Poplar Hill Dairy started out as a hobby—and, in Vince's Maefksy's words, "went wild," first starting out in this old dairy barn and then expanding to a modern, world-class goat operation. *Jen Brown*

She knew how to look after the goats as well as anyone, and Little Swan and Bear would follow her like two faithful dogs, and give a loud bleat of pleasure when they heard her voice.
— Johanna Spyri, *Heidi*

As a city girl, I'm not sure where my love of goats began. It might have been the eager friendliness of petting-zoo goats or early exposure to the book *Heidi* by Johanna Spyri. The antics of Little Swan and Little Bear on the Swiss mountain have fascinated many children and adults over the years. I was also drawn to the comfortable size of goats compared to cows or horses. Whatever the inspiration, when my husband, Wayne, and I moved to the country, my first hobby-farm animal was a goat.

Before goats took over our life, Wayne wondered what he'd agreed to as I started planning to have "a couple

goats, a calf, some pigs, and chickens." A few years later, he knew that his role as "critter control" had completely failed! The number of goats had increased exponentially. So we did what any sensible part-time farmers would do—we put in a commercial milking parlor. Our goat enterprise is called Terrapin Acres.

The reasons people keep goats are as widespread as the places goats are kept—and goat owners are as varied as the animals they raise. Sometimes called the "poor man's cow," a dairy goat can comfortably supply a family with milk while being kept on a fraction of the space and resources a cow requires. This makes the goat an ideal animal for the back-to-the-land movement that surges every few decades. Goats are therefore the animal of choice for home-school families living on a small acreage.

Goat projects also teach children the responsibility of caring for a living creature. Easy to handle, gentle,

Working in a pit milking forty goats at a time was an education. *Jen Brown, Poplar Hill*

Young people aren't the only ones who can benefit from keeping goats, though. As the United States population becomes less rural, adults as well as children have less contact with the land and the animals raised in the country. Learning about the cycle of life by owning a farm animal is one of the most enriching experiences imaginable.

Goats have their practical uses, too. I keep goats as a multipurpose animal, and I try to use all the products my goats provide. I eat their meat and drink their milk. Some of my friends won't touch goat meat because they feel that doing so would be like eating a pet. I must admit that some of my own goats never reach the freezer for this very reason!

This book is a compilation of my wanderings and an introduction to the caprine world. If you have an interest in raising goats, I hope to give you an idea of what to expect—and how to get help when your capricious goats don't do what you expect. There are as many ways to raise and care for goats as there are goat keepers. Enjoy your journey!

and gregarious, goats make a great introduction to raising livestock. FFA and 4-H goat projects are popular in many parts of the country, and I have met some bright and caring young people during my years showing and breeding goats. A number of these kids previously had little or no experience with livestock.

Terrapin Acres is the name of my goat farm in Scandia, Minnesota. Our first goats were Nubian dairy goats. *Terrapin Acres*

CHAPTER 1

• • • • • • • • • • • • • • •

A GOAT IS JUST A GOAT — RIGHT?

People are goats, they just don't know it.
—Ches McCartney, the Goat Man

When I decided to raise goats, there were just a few recognized dairy breeds. The LaManchas, with their funny little ears, were easy for me to rule out—they looked too goofy. The Swiss breeds were nice, just not what I wanted. Since I wanted a dairy goat, I didn't consider the tiny Pygmy or the silky-coated Angora. The heavily muscled Boer goat hadn't reached this country yet, and the stiff-legged Tennessee fainting goat wasn't even on my radar. The Nubian, with her long ears and variety of colors, drew my eye. Just like that, my choice of goat breed was decided, and a journey I'd never anticipated began!

I'm far from alone in my fascination with goats. Geographically, goats are the most widespread livestock species. Because of their adaptability, goat herds can be found anywhere from the cold mountains of Siberia to the deserts of Africa to the moist regions of the tropics.

Scientists think that the wild bezoar (*Capra aegagrus*), an ancestor to the goat that still thrives in Europe and Asia, was the first domesticated herbivore. In the Fertile Crescent of the Middle East, archaeological studies have found evidence that goats were bred in early human settlements.

Other studies suggest that Asia was another center of caprine development. Through DNA analysis, researchers have found a completely different strain of Asian goats, along with unique strains of pigs, cows, and sheep. Wild goats in Pakistan may have been the ancestors of Cashmere goats. Many traits of the Cashmere are different from the rest of the goat breeds.

Despite these multiple origins, animal geneticists have concluded that the world goat population has less genetic variation than the world cattle population. Goat DNA shows only about a 10-percent variation between animals on different continents. Cattle DNA varies by 50 percent or more. This lack of variation seems to indicate that goats migrated with their keepers across continents.

Migratory people likely took along the most versatile animals. Goats provided milk, meat, fiber, and hides while also carrying some of the load on their backs or pulling carts of gear. Their adaptability and their calm, friendly nature make goats easy to transport. In colonial times, they traveled on the *Mayflower* and other ships to provide food. These goats stayed with colonists or were sold as trade goods. Some were released on islands to run wild and provide supplies for sailors on future voyages—not a great idea. In the absence of natural predators, the non-native goats soon overran their island homes.

Interest in goats is growing. A 2007 inventory by the United States Department of Agriculture (USDA) totaled almost 3.6 million head, 4 percent higher than in 2006. Included were 2.69 million breeding goats, 335,000 dairy goats, 260,000 Angora goats, and 3 million meat goats. Since these numbers come from agriculture census data that does not survey pets or hobby farm animals, it is impossible to guess how many goats actually live in the United States!

One likely reason for the rising numbers of goats is that ethnic populations are booming in the United States, and many of these communities want traditional foods from their native lands. Others are simply interested in food from sustainable sources.

Goat horn has traditionally been used for musical instruments such as the Jewish shofar. *Shutterstock*

The use of goats and their products is limited mostly by public perception. Some societies prize goats; others view goat products with suspicion. Goats make excellent companions and beasts of burden but are better known for the products they provide, which vary as widely as the animals they come from. The obvious include cheese, milk, hair, and meat. But goat hide, a byproduct of the meat goat industry, has many uses. Goatskin parchment was once used for writing in Europe. Wine and water were carried in containers made from goat hide. A soft, fine goatskin called morocco is imported from Africa today and used by modern bookbinders because it is strong and durable. Musical instruments, especially drums, are made from goat hide. Goat leather also makes fine boots and clothing. The phrase "kidskin gloves" still brings to mind the finest of soft leather. In fact, my favorite gardening gloves are made from goatskin!

Even the horns of goats have numerous uses. The modern flutophone probably originated from the *gemshorn* (German for "goat horn"), a medieval instrument played like a flute and made from goat horn. The shofar, a ceremonial horn used in Jewish tradition, is sometimes made of goat horn. Goat horn has been used to make eating utensils and drinking containers. Goat horn is still used today for jewelry, buttons, and crafts.

The range of color and form found in goats is amazing! *Jen Brown, Terrapin Acres*

Census figures show 335,000 dairy goats in the United States.
Jen Brown, Poplar Hill

TYPES AND BREEDS OF GOATS

In spite of the genetic similarities among the goat population as a whole, experts have identified more than three hundred distinct caprine breeds. A breed is defined as a group of animals that has certain traits in common, such as similar color, conformation, function, or size. Animals within a breed pass these traits to their offspring. Landrace breeds are those that evolved in the wild through natural selection and are ideally suited to their specific environment, while modern domestic breeds are created through controlled matings.

A wide variety of organizations represent goats by type or breed. Breed associations create criteria for maintaining specific breeds.

DAIRY GOATS

The American Milch Goat Record Association began registering dairy goats in 1904. (*Milch* derives from an Old English word meaning "giving milk.") Now known as the American Dairy Goat Association (ADGA), this venerable organization has more than 8,000 regular members and nearly 5,000 junior members.

The American Goat Society (AGS) was founded in 1936 by breeders wanting to maintain the purebred status of their pedigreed dairy goats. The next year, AGS merged with the International Dairy Goat Record Association, previously established in 1925. While some registries allow non-pedigreed stock to gain purebred status via breeding with purebred goats, AGS has no program for these crossbred, or "Grade," goats.

In ADGA, however, when two different breeds of registered "Purebred" goats mate, their offspring are eligible to be recorded as Grade or Experimental. Purebreds that have serious enough defects to disqualify them from being registered in their breed must also be recorded as Experimental. For example, Sable Saanens were listed as Experimental before being recognized as a separate breed from traditional white Saanens.

Another difference between ADGA and AGS is that ADGA permits the registration of goats designated by breed name with the addition of "American." A Purebred dairy goat comes from a Purebred sire and Purebred dam of the same breed conforming to breed standards. An American goat is the offspring of a sire and dam of the same breed going back a minimum three generations for does and four for bucks. ADGA maintains separate herd—official lists of registered animals—for Purebred and American goats. The LaMancha and Sable breeds have an open herd book, meaning that goats can "breed up" into the Purebred registry. Nigerian Dwarf goats have only a Purebred herd book with no Grade program.

Dairy goats are distinguished by ear type, color pattern, and size. The Swiss breeds (Alpine, Oberhasli, Saanen, Sable, and Toggenburg) have upright ears. These breeds are distinguished from each other by color.

Alpine Goat Pattern

Alpine coat patterns are expressed in French terms. When a recognized coat pattern is banded or splashed with white or another color, it is described as broken, as in "broken chamoisee."

- **chamoisee**. Brown or bay markings with a black face, dorsal stripe, feet, legs, and sometimes a martingale running over the withers and down to the chest. A **two-tone chamoisee** goat has light front quarters with brown or gray hindquarters. (The male version is **chamoise**.)

- **cou blanc** ("white neck"). White front quarters and black hindquarters with black on the head.
- **cou clair** ("clear neck"). Tan front quarters and black hindquarters.
- **cou noir** ("black neck"). Black front quarters and white hindquarters.
- **pied**. Spotted or mottled.
- **sundgau**. Black with white markings on underbody and face.

Jen Brown

The Alpine is the second most popular dairy goat registered by the American Dairy Goat Association. The ideal Alpine has a straight face; a Roman nose is faulted. *Jen Brown*

Alpine

The ancestors of all purebred Alpines in the United States today arrived in 1922 when twenty-one animals were imported from France. Alpine goats are sometimes called French Alpines.

The Alpine has medium to short hair and upright ears. The breed standard requires a straight face; a Roman nose (convex muzzle) is faulted. Toggenburg coloring (brown body and white markings) or all-white coloring is discriminated against. Mature does should stand 30 inches tall at the withers and weigh 135 pounds. Mature bucks should stand 32 inches tall at the withers and weigh 160 pounds.

LaMancha

The name *LaMancha* originated from an unreadable description of some short-eared goats sent for exhibition to the 1904 Paris World's Fair. The name was illegible, but the words "La Mancha, Cordoba, Spain" were readable. This breed was probably originally known as Murciana.

The "gopher ear" style on LaManchas measures less than one inch with little or no cartilage. This is the only type of ear eligible for registration of LaMancha bucks. *Jen Brown*

In 1960, so-called "elf-ears" were prohibited on registered LaMancha bucks. The "elf ear" must be less than two inches long with the end of the ear turned up or down. Some cartilage shapes this small ear. *Jen Brown*

Arriving in California with Spanish missionaries, these short-eared, dual-purpose milk-and-meat goats spread through the American West. In the 1920s, Phoebe Wilhelm crossed about 125 descendants of the mission goats with Toggenburg bucks. Later, Alpines, Nubians, and some Saanen bucks were bred to the short-eared does. Eula Fay Frey of Oregon worked hard to get LaManchas accepted for registry. In 1958, ADGA registered the first LaMancha—Fay's Ernie, L-1. About two hundred animals, sixty of which belonged to Frey's herd, started the registered herd book. The small ears are a distinctive breed characteristic and a dominant trait. Short ears carry even when crossed with another breed.

I didn't consider LaManchas for my first goats because of their appearance. This common initial reaction to the pixie-eared goat is quickly overcome by personal experience. The gentle, curious personality of the LaMancha wins people's hearts.

Nigerian Dwarf

Zoos initially brought miniature goats to the United States to feed large cats. The gentle nature of minis led to their popularity as pets. Nigerian Dwarfs and the Pygmy breed (described below) share the same genetic base, but over time breeder selection split them into two distinct breeds. Once considered at risk, Nigerians have benefited from increases in their popularity as a dairy goat. The American Livestock Breeds Conservancy (ALBC) now designates Nigerians as recovering on their Conservation Priority List.

A dependable dairy goat, LaManchas milk for a long, steady lactation, giving high volume, butterfat, and protein. They are easy keepers.
Jen Brown

13

The Nigerian Dwarf is a miniature breed of dairy goat. ADGA standards require does stand no more than 22.5 inches (57 cm) and bucks no more than 23.5 inches (60 cm). This height was established after years of wrangling by breeders. *Jen Brown, Cutter Farms*

The head, limbs, and body of a Nigerian goat are proportionate, a condition known as pituitary dwarfism. In 1981, the AGS was the first registry to recognize the Nigerian as a dairy goat. The International Dairy Goat Registry (IDGR) started recording the breed in 1982. The breed was accepted into the ADGA registry in 2005. Before these dates, Nigerian goats were considered solely a pet breed.

Kathleen Clapps of Texas was the first breeder to enter Nigerian Dwarf goats into an official milk test. Her goat earned Advanced Registry Star Milker status with 427 pounds of milk, 25 pounds of fat, and 20 pounds of protein. This level of production, not out of the ordinary for a traditional dairy goat, showed that Nigerians could successfully meet the standards set by the larger breeds.

Nigerian Dwarfs enjoy people and can become attached to their owner. Of course, these traits may lead to a "talkative" goat that lets you know when she wants something! Because of their small size, Nigerians are a favorite with 4-H families.

The original Nigerian-type goats from Africa were black, a recessive color. Crossbreeding established the breed and gave today's Nigerian Dwarf a variety of colors and patterns. The Nigerian breed standard set by the ADGA includes short, fine hair; a straight or dished face; and erect, alert ears of medium length.

Nubian

The Nubian is a combination of English goats and goats from other parts of the world. English ships traveled to many parts of the world, carrying goats to provide fresh milk. Goats from foreign ports were crossed with common English milking goats.

During the 1800s, some goats were brought to France from Nubia, a region in North Africa. Descendents of these goats, called Nubians, arrived in England in 1883. Recognition for this long-eared breed came quickly. By 1896, the Nubian was a registered breed in England.

Nubian-type animals came into America as early as 1896; however, those bloodlines have disappeared. The beginnings of the first registered Nubians in America are traced to animals imported by J. R. Gregg of California. By 1918, forty animals were registered as purebred Nubians in the United States. It was a quiet beginning for the breed that has become the most popular dairy goat in the United States. Nubians outnumber all other currently recognized dairy breeds two to one.

The Nubian is a dual-purpose goat, useful for both meat and milk. Averaging less milk than the Swiss breeds, Nubian milk has high average butterfat content, between 4 and 6 percent. The Nubian breeding season is longer than that of the Swiss breeds, so it is easier to breed Nubians for off-season milk.

With her long, graceful ears, a proud Roman nose, and a variety of colors, the Nubian is the most popular breed of dairy goat in the United States. *Jen Brown, Terrapin Acres*

Mid-sized, vigorous, and alert, Oberhaslis are described as chamoise in color. Chamoise is a rich red color, preferred over lighter tones or black. Bucks often have more black on the head than does plus more white hairs. *Susanna Yoemans, Cardinal Hill*

According to the breed standard, a mature Nubian doe should stand 30 inches at the withers and weigh 135 pounds. A male should stand 35 inches at the withers and weigh 175 pounds. Nubian hair is short, fine, and glossy. Any colors or patterns are acceptable.

Oberhasli

Beginning in the 1930s, an Alpine-type goat was imported into the United States from the Oberhasli region of Switzerland. The goat was originally known as the Swiss Alpine and considered a color variety of the Alpine breed. In 1979, the ADGA recognized the Oberhasli as a separate breed. The Oberhasli Herd Book began in 1980.

Like the Nigerian Dwarf, the Oberhasli was once listed as endangered by the ALBC. Today, more breeders are raising this attractive animal, which is now increasing in number and designated as recovering.

Saanen

Saanens get their name from the Saanen Valley in the south of Switzerland. Between 1904 and the 1930s, about 150 Saanens were imported into the United States. These, in addition to later imported animals from England, formed the foundation of purebred Saanens in the United States.

Majestic white Saanens are a popular dairy animal because of their calm, eager-to-please temperament. Known as the Holsteins of the goat world, Saanens produce, on average, the most milk of any of the dairy breeds.

The Saanen is the largest of the dairy breeds with rugged bone and plenty of vigor. White color is preferred, although light cream is acceptable. *Jen Brown, Poplar Hill*

Sable

Established as a separate breed by the ADGA in 2005, the Sable is essentially a colored Saanen. Due to a recessive gene in the original white Saanens, colored goats have always been a part of the Saanen bloodlines.

The first Saanens were registered regardless of their color. In the late 1930s, the Saanen Breeders Association adopted a resolution that future Saanens had to be white to be acceptable for registry.

The Sable Herd Book is an open herd book like the LaMancha's. After three generations of American Sables in the pedigree, the next generation that meets breed standard is eligible for Purebred Sable status. A first-generation American Sable may have two Purebred Saanens as parents but meet the Sable requirements strictly based on color.

The reverse is not true. If two Purebred Sables produce an all-white offspring, that kid is now only eligible to be recorded as Experimental since the color standard for Sable specifies anything "except solid white or solid cream." In order for color to show in the offspring, both the sire and the dam have to pass the color gene to the kid.

Toggenburg

The Toggenburg is the aristocrat of the dairy-goat community. Among the first purebred dairy goats to come to the United States, these animals had impressive family trees back home in the Toggenburg Valley of Switzerland. Swiss exporters in the early 1900s proudly claimed the breed had been pure for three hundred years. The Swiss Toggenburg breeders association calls the Togg "the oldest and purest breed in Switzerland."

At the turn of the previous century, these chocolate-colored goats with striking white trim were the most common dairy goat found in the United States. A famous herd of Toggenburgs was raised in North Carolina by Lillian Sandburg. (Her husband, Carl, was famous for his poetry, while she was well known in dairy-goat circles.) A case can be made that the modern dairy-goat industry was founded on the Toggenburg. This may come as a surprise to modern dairy-goat lovers, as the Toggenburg has slipped to the fewest number of animals registered of any of the dairy breeds.

Sometimes called deer-like in stance and appearance, the Toggenburg has longer hair than the other dairy breeds. The coat can be shorter in America because no emphasis has been placed on hair length. In Switzerland and England, long hair is desirable on the shoulders and back legs.

Size is another breed characteristic that changed as the animal moved from its native country. Swiss and British purebred Toggs are small, economical animals. Meanwhile, in the United States, Toggenburgs have increased in size. A few years ago, ADGA standards changed from 125 pounds and up to 115 to 150 pounds.

The breed has won many best udder awards. According to the Dairy Herd Information Association (DHIA), the all-time milk producer for all breeds is a Togg doe that produced 7,965 pounds of milk in 305 days. Many Toggs milk into their early teens.

A white pattern of markings is specific to Toggenburgs, including erect white ears carried forward with a dark spot in the center and white stripes running from above each eye down to the muzzle. *Jen Brown*

FIBER GOATS

The most common fiber goat in the United States is the Angora, with 238,000 tallied by the National Agricultural Statistics Services All-Goat Survey in 2007. The hair of the Angora goat is made into mohair, and the United States is one of the largest suppliers of mohair in the world. Cashmere goats are less numerous. Crosses are popular with hobbyists and hand-spinners.

Angora

Angora goats came to the United States in 1849 under noble circumstances. Dr. James Davis of South Carolina helped introduce cotton production to Turkey. The Sultan reciprocated, sending Dr. Davis a gift of mohair-producing animals. Purebred Angora goats had arrived eleven years earlier in South Africa. Both South Africa and the United States dominate present-day mohair production.

The only U.S. registry for purebred Angoras is kept by the American Angora Goat Breeders Association (AAGBA), established in 1900.

Cashmere

Adult Cashmere goats are sheared once a year and yield as much as 2.5 pounds of fleece. The hair on Cashmere goats comes in two types: harsh guard hairs and fine under wool. The fleece must be dehaired to remove the guard hairs from the luxury fibers. A fleece cleans out at 40 to 60 percent cashmere.

There is no specific breed registry for Cashmere goats. Cashmere wool is grown by all caprine species except the Angora. A Cashmere goat is one that produces under-down of commercially acceptable color and length. Cashmere goats are judged by the quality and quantity of the under-down and the size or build of the animal. The quantity and quality of cashmere fiber produced are determined by the crimp (or style), the diameter and length of the fiber, and how much of the animal's body is covered with down.

Cashmeres are generally raised as dual-purpose animals for fiber and meat. The goats have been bred to have wide horns, blocky builds, and refined features. Because their feral origins are more recent than those of other breeds, Cashmere goats tend to be wary rather than

The Angora goat is gaining popularity across the country with hand-spinners and hobby breeders. These goats come in all different colors with white a dominant color. *Shutterstock*

A Cashmere goat produces soft, high-quality fiber, prized for sweaters and other garments. *Shutterstock*

placid. Breeders report that they are easy kidders and good mothers.

A cross between an Angora and a Cashmere goat is called a Cashgora. Its coat has more of a crimp than cashmere and fewer guard hairs. The fiber is said to combine cashmere softness with angora curl and to take dye nicely.

Colored Angora Goat

Until recently, Angora goats were bred solely for white mohair. Breeding Colored Angora goats is challenging, but the naturally colored mohair is popular with hand-spinners. In 1999, the Colored Angora Goat Breeders Association was formed.

The two classifications of Colored Angoras are Blue Card and Red Card. Blue Card goats have a distinctive, predominantly colored fleece. Red Card goats may have a white or lightly colored fleece.

Nigora

Crossing small-breed goats with their large-breed counterparts has become popular. A cross between the Nigerian Dwarf and the Angora, the Nigora is a miniature fiber breed. These goats come in a variety of colors. They may have mohair or cashmere fiber, but the typical fleece falls into the category of cashgora. A breed association was formed in 2007.

Pygora

The Pygora is another small-breed cross. Angora does were bred to Pygmy (described below) bucks. The short, fuzzy Pygoras are popular with hand-spinners, hobbyist breeders, and pet owners.

The Pygora is essentially a Pygmy Angora goat, possessing the fine hair coat of the Angora and the small stature of the Pygmy. *Fran Bishop, Rainbow Spring Acres, Pygora Breeders Association*

The Pygora Breeders Association, formed in 1987, has seen an increase in the popularity of its breed. Breed standards require that the registered Pygora goat be no more than 75 percent registered Angora goat or 75 percent registered Pygmy goat. The offspring of an Angora and a Pygmy is considered not a Pygora but simply a cross; these animals are registered as first generation.

MEAT GOATS

Meat goats are the most numerous type of goat in the United States, with 2.4 million head in 2007. Most meat-goat herds are located in the Southwest and South, but numbers are increasing across the country as goat producers struggle to meet a rising demand that far exceeds their capacity.

Boer

Boer goats were originally developed as a meat breed in South Africa. *Boer* means "farmer," the same name given to Dutch farmers in South Africa. This farmer's goat is known for heavy muscling and rapid weight gain.

Boers initially came to the United States by way of Australasia, as the USDA restricted the import of goats directly from Africa. In the late 1980s, South African producers smuggled Angora embryos into New Zealand and Australia; almost as an afterthought, some Boer embryos were included. Two main farms, African Goat Flocks and Landcorp, then exported the very first Boers to the United States. Those first Boers arrived in 1993 from New Zealand at export costs of $8,000 to $10,000. What followed was a frenzy of exotic and speculator buying. One Boer buck sold for $80,000!

When I attended the first American Boer Goat Association national convention in 1994, it surprised me that people could get so worked up about a goat. The bubble burst quickly. By mid-1995, the USDA had changed import rules to allow goats and embryos to come directly from South Africa. That same year, some auctions were selling does for as little as $800. Some Boer breeders worry that the initial pampering of these rare, expensive animals affected their hardiness and growth. Today, Boers of acceptable quality can be purchased at the same cost as other goat breeds.

The Boer goat is a popular meat breed. *Jen Brown, Cutter Farms*

Confusion exists about the definition of a true full-blooded Boer goat. Regardless of the origin of import, all Boer goats tracing their pedigree directly back to South Africa are full-bloods. Percentage Boers are those with at least one goat of non-African breeding in the pedigree.

Kiko

The Kiko was bred by a New Zealand consortium of large goat farms. This group crossed feral does with Anglo-Nubian, Toggenburg, and Saanen bucks. Within four generations, a new breed of meat goat was born. It showed dramatic improvements in live weight. The new meat breed was christened the Kiko (for *kikokiko*, Polynesian for "flesh consumption"). By 1986, New Zealand breeders had closed the Kiko herd book. It is the only goat bred specifically for performance rather than appearance.

Efforts are being made to breed the Kiko true to its hardy origins. Known for rapid weight gain and foraging ability without supplemental feed, the Kiko is supposed to thrive under all conditions without human intervention.

Most Kiko goats are white with brown eyes and dark skin color, although the American Kiko Goat Association allows goats with other skin, hair, and eye colors to be registered.

Myotonic Goat

Myotonic, Tennessee Fainting, Tennessee Meat, Texas Wooden Leg, Stiff, Nervous, and even Scare—these names all refer to a goat breed with a condition called myotonia congenita. The muscle cells tighten when the

All fainting goats are heavily muscled because of the extra work these muscles do as they stiffen. Short and long-haired varieties exist with some producing cashmere. *Jen Brown, Hidden Springs Game Farm*

The concern about losing genetic variation among livestock is real. Rare breeds carry traits for disease resistance, mothering ability, and weather tolerance that can be lost in commercial herds bred for conformity and production. Problems also occur when commercial demands spread the animals to environments less suitable than their native climates. Uniformity of type may be convenient for large-scale farming, but it creates vulnerabilities. In a worst-case scenario, a dominant commercial breed could catch a deadly disease. Such a calamity could wipe out the group.

The American Livestock Breeds Conservancy works to conserve historic breeds while supporting genetic diversity in livestock. This group has six goat breeds on its Conservation Priority List. The two rarest are the Arapawa and the San Clemente, landrace goats that developed in the isolation of island homes after being deposited there during colonial expansion from Spain and England. Other listed breeds are the Myotonic, the Oberhasli, the Nigerian Dwarf, and the Golden Guernsey.

Arapawa

Olde English Milche goats arrived in the South Pacific with Captain James Cook and other colonial explorers in the late eighteenth century. One such location was Arapawa Island in the Marlborough Sounds of New Zealand, where a group of feral goats flourished for 150 years. Now extinct in their native England, these goats recently faced extermination at the hands of New Zealand authorities, who considered the herd nonnative vermin. The controversy caught the attention of Betty and Walter Rowe, who had moved to New Zealand from the United States in the 1970s. Their efforts led to the formation of the Arapawa Wildlife Sanctuary in 1986, which continues to preserve the breed today.

The ALBC has listed the Arapawa as a breed to study since 2004. Recent studies conducted by Dr. Phillip Sponenburg of the ALBC showed the Arapawa to be a unique genetic breed. According to the Arapawa Goat Breeders–USA, the registered total in the United States is merely 175 animals.

Golden Guernsey

A minority breed even in England, the Golden Guernsey has long been coveted by American goat fanciers. The import of these animals is presently impossible. Guernsey-type goats are being bred in the United States using Golden Guernsey semen crossed with Swiss dairy goats. The only true Golden Guernsey, however, is one registered in the British Golden Guernsey Herd Book.

A smaller breed, Arapawa goats are lean and have light-boned frames. They vary in color, although most registered animals are white, gray, or black mixes. *Al Caldwell, Long Path Farm*

Golden-haired dairy goats are mentioned in English records as early as 1826. The goats initially arrived with trading ships. The Golden Guernsey lines most likely came from the Maltese Islands with origins in Greece and Syria. Starting in the 1920s, they were registered in the English Guernsey Herd Book. By 1965, a distinct group of golden goats was breeding true, and a separate Golden Guernsey Herd Book was started. This closed registry allows no grading up from crosses with other breeds.

Southwind Farms in New York has the only herd of purebred Golden Guernseys in the United States. Their first embryo-offspring were born in 1998 after the import of embryos through Canada a few years earlier.

San Clemente

Off the coast of southern California lies San Clemente Island, home to a landrace breed of goats that inhabited the island following visits by colonial ships. Unique genetic markers show these goats to be a distinct species.

The U.S. Navy has been responsible for San Clemente Island since 1934. Locked onto the island, the San Clemente goat population soared. By a 1972 survey, at least fifteen thousand goats were living on 57 square miles. The goats destroyed vegetation and endangered the ecosystem. In the 1970s, San Clemente Island was made a preserve for native species. Since the goats were not native, they had to go.

The Navy first used hunting and trapping to eliminate the feral goats. The animal protection group Fund for Animals spearheaded a program of systematic removal, which saved certain animals from destruction. Bucks were neutered and females adopted out with "no breed" clauses. No goats remain on San Clemente Island.

The San Clemente Island Goat Association is working to increase the geographical distribution and preserve the genetic diversity of San Clemente goats. Only about two hundred of these animals remain today.

Golden Guernseys come in any shade of gold and may have some white marking. They are smaller in size and less wedge-shaped than other dairy goats. *Phyllis Clayton, Golden Guernsey Goat Society*

A PHYSICAL AND
BEHAVIORAL OVERVIEW

Once you learn the many parts of a goat, you'll be able to master the quiz bowls put on by 4-H and the National FFA Organization. There are also more practical reasons for goat breeders to be familiar with these terms. I have learned quite a bit about good conformation by watching judges or appraisers evaluate animals. Until I learned the language, their comments meant very little. Now I can appreciate a goat that is "higher in the chine" or "wider in the escutcheon." By studying the charts that follow, you can also learn to evaluate your goat.

If you put a silk dress on a goat, he is a goat still.

—Irish proverb

PARTS OF THE GOAT

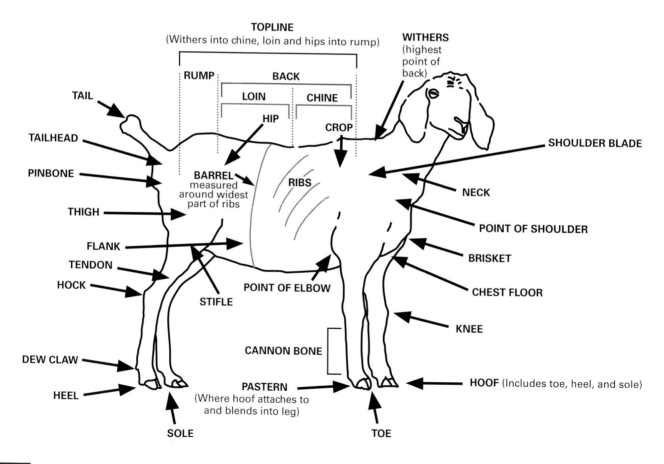

TOPLINE
(Withers into chine, loin and hips into rump)

WITHERS
(highest point of back)

RUMP · BACK
LOIN · CHINE
HIP
CROP

TAIL
TAILHEAD
PINBONE
THIGH
FLANK
TENDON
HOCK
STIFLE

BARREL
measured around widest part of ribs

RIBS

POINT OF ELBOW

SHOULDER BLADE
NECK
POINT OF SHOULDER
BRISKET
CHEST FLOOR
KNEE

DEW CLAW
HEEL
SOLE

CANNON BONE

PASTERN
(Where hoof attaches to and blends into leg)

TOE

HOOF (Includes toe, heel, and sole)

PARTS OF THE MAMMARY SYSTEM

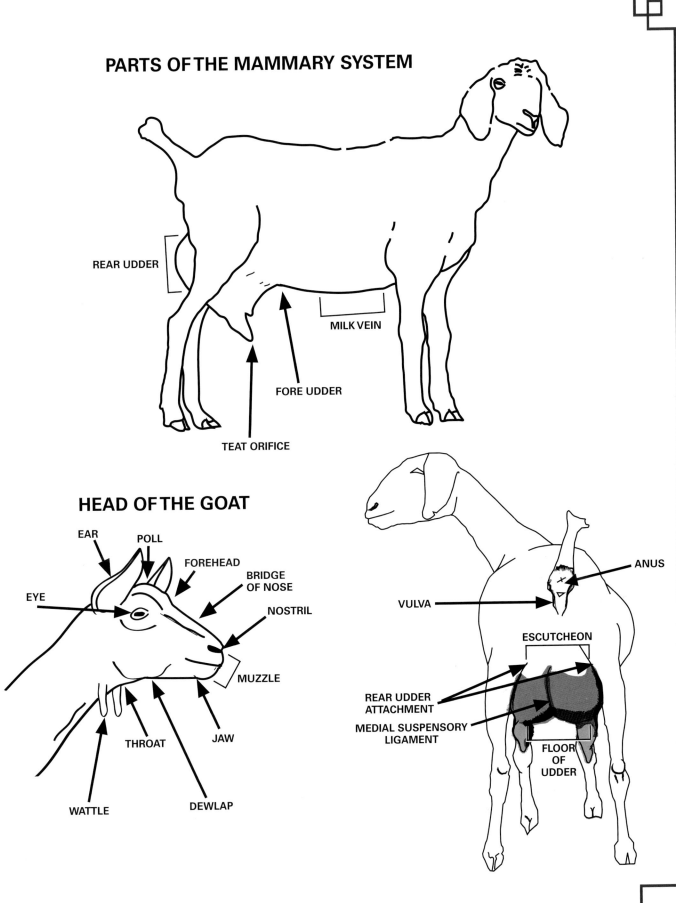

REAR UDDER

MILK VEIN

FORE UDDER

TEAT ORIFICE

HEAD OF THE GOAT

EAR

POLL

FOREHEAD

BRIDGE OF NOSE

NOSTRIL

EYE

MUZZLE

THROAT

JAW

WATTLE

DEWLAP

ANUS

VULVA

ESCUTCHEON

REAR UDDER ATTACHMENT

MEDIAL SUSPENSORY LIGAMENT

FLOOR OF UDDER

PARTS OF THE BUCK

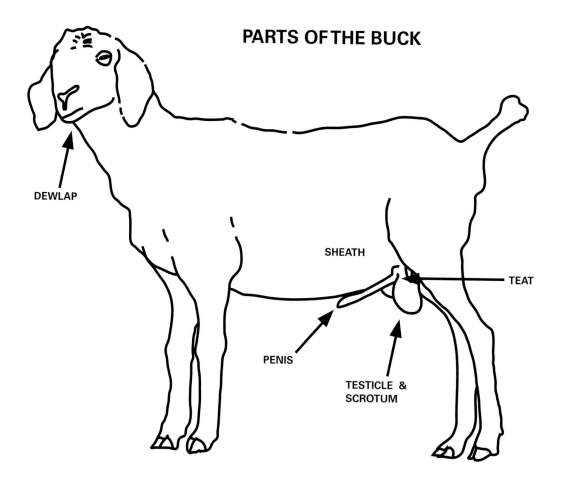

DEWLAP

SHEATH

TEAT

PENIS

TESTICLE & SCROTUM

The rectangular shape of the pupil gives goat eyes a unique appearance. Most goats have brown eyes (*above*), although some have blue eyes (*right*). *Jen Brown*

Goats, like cows, raise and lower themselves by bending their front legs first rather than their back legs. *Jen Brown*

MOUTH

The goat's mouth has both a hard palate and teeth. There are no teeth on the top front. When goats browse, they use the bottom teeth and the hard upper gum pad to break off vegetation.

Just because you don't see opposing teeth doesn't mean there aren't any. In the back of the goat's mouth are grinding teeth. Any child who has managed to stick his hand far enough into a goat's mouth will tell you that these teeth can hurt! The goat passes food by the tongue into the back of the mouth, where the opposing teeth grind it into pieces.

It is important that the teeth meet the palate correctly in the mouth. Two teeth-related congenital conditions can be found in goats. Parrot mouth occurs when the top jaw is longer than the bottom jaw. The other condition, undershot or monkey mouth, occurs when the lower jaw is longer than the upper jaw. Either condition is grounds for culling, as each makes it more difficult for the goat to collect and chew browse, which can reduce feeding efficiency and milk production.

The front teeth can be a good indicator of a goat's age. At birth, there are six incisors on the bottom jaw. Early in the goat's life, these teeth are just starting to break through the skin. By four weeks, there are normally eight incisors known as milk teeth, in the front of the lower jaw, as well as twelve molars in the back of both the top and bottom jaws. The teeth wear with age, so the older the goat, the more worn down the teeth become. The teeth also spread in the mouth, become loose, or fall out. A goat that has lost all its teeth is known as gummy.

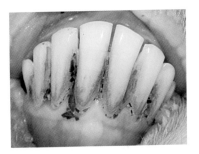

The upper jaw of the goat's mouth has no front teeth. Each year, baby teeth on the bottom jaw are replaced by larger adult teeth. By the age of four years, all eight incisors are usually mature and the goat is now known as an "8 Tooth." *Jen Brown*

27

DIGESTIVE SYSTEM

As a ruminant, the goat has a stomach system made up of four compartments. The first two chambers are the rumen and the reticulum (also known as the honeycomb), which work with saliva and stomach juices to break down fiber. The final two chambers are the omasum and abomasum, or true stomach.

In kids, milk bypasses the other chambers to be digested directly by the abomasum. Stomach acids in this final chamber digest the milk without the need for fermentation to break down roughage. At this stage of development, the abomasum is the largest chamber. As the kid starts eating roughage, the rumen and other chambers become more active. By adulthood, the rumen has grown to be the largest chamber.

At fairs, I am often asked if one of my older milkers is pregnant. I explain that the large, round belly is caused by her rumen. This big fermentation chamber allows the ruminant to break down cellulose into digestible materials with the help of enzymes and bacteria. The goat chews roughage and swallows it into the rumen. Once it is mixed with digestive juices, the mixture is regurgitated in the form of cud. Cud is chewed and swallowed multiple times and passed between the rumen and reticulum until it is semi-liquid.

The finely chewed product is then swallowed again and passes into the next stomach chamber—the omasum. The omasum is where much of the water and mineral nutrients are absorbed into the bloodstream. Finally, the semi-processed food passes into the abomasum. There, additional digestion takes place before the contents pass into the intestinal system.

BODY CONDITION SCORING

Body condition scoring is a system used to evaluate the amount of body fat on an animal. The score indicates whether the goat is skinny, just right, or fat. A 1 to 5 classification system is the most common, with 1 being very thin and 5 being very fat. Most goats thrive in a body condition score 3.

Cud is chewed using the back teeth. *Jen Brown*

The American Dairy Goat Association has a formal process for evaluating dairy goats known as Linear Appraisal. Senior dairy goat judges trained as evaluators come to your farm and assess the animals using a standardized format that gives breeders a measure for breeding goats approaching a certain type. Other appraisal systems are in place with various breed associations. *Carol Amundson, Terrapin Acres*

Body Condition Scoring

SCORE	CONDITION	BACKBONE AND RIBS	LOIN
1	Very lean	Easy to see and feel Can feel under ribs	No fat
2	Lean	Easy to feel Smooth Need to use a little pressure to feel ribs	Smooth fat
3	Good	Smooth and rounded Even feel to the ribs	Smooth fat
4	Fat	Can feel backbone with firm pressure No points on spine and no ribs felt Indent between ribs felt with pressure	Thick fat
5	Obese	Smooth No individual vertebra felt No separation of ribs felt	Thick fat Lumpy Jiggles

This under-conditioned milker has very little fat on her body, earning a Body Condition Score of 1 to 2. *Jen Brown*

This over-conditioned wether has quite a bit of fat under his winter coat, earning a Body Condition Score of 4 to 5.

A playful goat bounds and leaps and quivers. *Barb O'Meehan*

Prayer of the Goat
        by Carmen Bernos de Gasztold

Lord, let me live as I will!
A little giddiness of heart,
the strange taste of unknown flowers.
For whom else are Your mountains?
Your snow wind? These springs?
The sheep do not understand.
They graze and graze,
all of them, and always in the same direction,
and then eternally
chew the cud of their insipid routine.
But I—I love to bound to the heart of all
Your marvels,
leap Your chasms,
and, my mouth stuffed with intoxicating grasses,
quiver with an adventurer's delight
on the summit of the world!

BASIC GOAT BEHAVIOR

The images in Carmen Bernos de Gasztold's poem are right on the money! She emphasizes that goats are not sheep.

The layperson tends to lump sheep and goats together. Both species are ruminants. Both provide meat, milk, hair, and hide. But there are key differences. Sheep and goats don't eat the same way. Sheep graze, feeding almost exclusively on grass and plants growing on the ground. Goats browse, searching around for the most tender vegetation, wherever it might be. They nibble at the leaves from the bushes next to their pen, stand up on their hind legs to reach tree branches, and look to the choice new-grown blossoms of clover in their pasture. The false statement that goats eat anything—or, worse, that goats eat garbage—stems from their natural tendency to browse. Their inquisitive lips reach out to taste anything that catches the eye. The paper wrapping on a tin can is made from wood products. That paper, combined with the glue on the can, is very tasty. But eating it creates the impression that the goat wants to eat the tin can as well!

Goats are not as flock-oriented as sheep. Watch goats in the pasture. The herd spreads out, wandering. The animals go a set distance before they become too uncomfortable to separate farther from their herd mates. The lead goat is kept in sight, and that goat's actions influence the others. Sometimes, a small group of goats—or a loner—leaves the main herd to go adventuring. I have chuckled at the cry of distress and panicked running when a goat discovers that the herd has moved on while she had her nose buried in a bit of forage.

I recently picked up a "problem" goat. The owners were beside themselves because the solitary goat had destroyed their lilac bushes during her constant escapes. Penned with other goats, she settled right down. In the absence of other goats, a lone goat will look for company. The animal will escape its pen and cry plaintively. Then the stereotype of the goat in the clothesline and climbing on the car becomes reality.

Goats browse on a variety of plants. More like deer than sheep in their feeding habits, goats like to pick and choose their food.
Jen Brown

Another crucial difference between goats and sheep is their tolerance of the elements. Sheep have wool that repels water, allowing them to stay out in all sorts of weather. Goats hate to get wet. Even woolly Angora goats need protection from the elements. The lack of a water-repellant coat makes goats prone to chilling and pneumonia. It is easy to tell when rain is imminent, because the goats disappear from the pasture.

Although less flock-oriented than sheep, a lone goat often panics once it discovers it has become completely parted from the herd. *Barb O'Meehan*

Goats rear high in the air to confront attackers. *Jen Brown*

Differences Between Goats and Sheep

Goats

- Males are called bucks
- Females are called does
- Bearded
- Tail held upright
- 60 chromosomes
- Browsers
- Maintain some independence within a herd
- Susceptible to rain
- Butt downward after rearing on hind legs

Sheep

- Males are called rams
- Females are called ewes
- Beardless
- Tail hangs toward the ground
- 54 chromosomes
- Grazers
- Flock closely together
- Fleece protects from rain
- Butt charging straight ahead without rearing

Shutterstock

BEFORE BRINGING YOUR

GOATS HOME

Don't approach a goat from the front, a horse from the back, or a fool from any side.
—*Yiddish proverb*

Goats often prefer to stay near the barn, close to their source of food and water. *Jen Brown, Poplar Hill*

Goats don't require a lot of space unless you raise them on pasture or grow their feed yourself. A goat dairy on 5 acres might have more than one hundred milkers. Redwood Hills Dairy in California keeps over three hundred goats on 10 acres. Their facilities include open housing with little pasture beyond a lounging yard. Although this lifestyle may sound unappealing, goats actually prefer it. Even with access to outdoor space, dairy goats frequently lounge around the barn, enjoying the close proximity to water and feed. In fact, Redwood Hills is the first goat dairy in the United States to earn the Humane Farm Animal Care (HFAC) "Certified Humane Raised and Handled" label. Clearly, these goats are neither overcrowded nor abused.

Pet goats are usually kept in smaller numbers and housed accordingly. A large dog kennel can serve as caprine housing if the animal has plenty of exercise time with the owner.

Meat or fiber goat operations frequently run on range or pasture. Range requires the most land. Stocking rates of six to twelve goats per acre are recommended. The type and amount of forage available to the animals, as well as the weather, create many variables in the number of goats that a tract of land can support. A certain amount of feed and housing supplementation is usually required during harsh weather and during the breeding season.

REGULATIONS

You are more likely to encounter restrictions on your goat herd due to municipal zoning regulations rather than personal space limitations. Larger towns and cities have specific rules against livestock. Even in the country, subdivision covenants often regulate livestock.

Goats are often restricted to areas zoned as agricultural. Other regulations include acreage minimums, such as 5 acres minimum for keeping farm animals, and density limits, such as no more than ten mature goats per acre.

With mad cow disease, avian influenza, and other biosecurity concerns, premises identification is a hot issue for goat keepers. To prevent the spread of diseases, states often require health tests and animal identification. Some states do not allow livestock to be exhibited at shows unless the farm has a unique premises ID. Once your farm has this code, you must keep a record of all animals entering and leaving the premises. ID systems typically require all animals over a certain age to be permanently identified.

If you find yourself bringing your goats home from another state, be prepared. Interstate movement of animals can be tricky. Most states require a certificate of veterinary inspection, also known as a "health paper," for movement of goats across state lines. Testing for tuberculosis, brucellosis, bluetongue, and other livestock diseases may be needed. Animals from areas in which certain diseases are endemic, such as tuberculosis, may be prohibited from entering the state.

NATIONAL ANIMAL IDENTIFICATION SYSTEM

The National Animal Identification System (NAIS) is a proposed system designed to track specific types of animals and diseases. It was conceived by the U.S. Department of Agriculture, although state boards of animal health would likely be responsible for implementation should the program become mandatory. Various goat associations have taken a stand on this up-and-coming issue. Goat owners are lining up on both sides of the fence to either promote or oppose the system. Be aware of rules regarding animal ownership, movement, and distribution as they develop.

IDENTIFICATION

Permanent goat identification is used for registry and documentation purposes. The dairy-goat industry uses tattooing. The meat- and fiber-goat industries more commonly use ear tags. Other methods include microchip ID, ear notches, and freeze branding. In most instances, transported goats, sales (especially auctions), and exhibits require permanent ID.

Ear tags are required by some feedlots, even with the presence of tattoos or other ID. *Jen Brown*

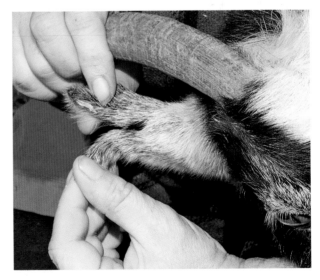

Goats can lose ear tags to a variety of factors, including thinness of the ear, browsing in areas that catch the tag, the curiosity of other goats, and infections. *Jen Brown*

TATTOOING

A tattoo is a series of numbers and letters inked into the ear or tail web of the goat. It is one of the most permanent forms of ID.

Your registry gives your herd identification letters; if you do not register, you may use your farm or personal initials. The herd ID goes in the right ear or on the right tail web, and the individual animal ID goes in the left ear or on the left tail web. For the left ear or tail web, a letter identifies the birth year of each kid born on your farm. The letters G, I, O, Q, and U are usually not used as birth letters, because they are hard to differentiate. Next to the year letter, put a number for the animal that uniquely identifies it from the rest of the year's crop. Most breeders start with 1 and work their way through the birth order. So if you've selected the letter "V" to stand for 2006, "V12" in the left ear of a dairy goat means it was the twelfth goat born in 2006.

MANAGEMENT SYSTEMS

There is a spectrum of methods for raising goats, each with its own advantages and disadvantages. Assess your

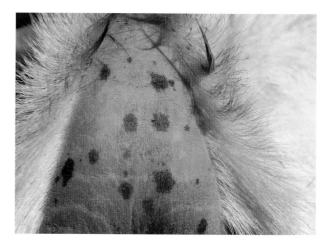

LaManchas are tattooed in the tail because their ears are too small. *Jen Brown*

own personality and resources in order to decide how to proceed with your herd.

Conventional caprine management techniques offer the widest number of options. Advantages include easy access to supplies and feeds, plenty of documentation and resources on techniques, and often lower supply costs. Many of these techniques were designed for large operations. Large herds may use a regular schedule of antibiotics, wormers, or chemicals to prevent diseases and pests. This routine may prevent health problems before they start brewing.

Disadvantages of conventional management exist in the same areas that make it advantageous. Routine use of antibiotics and wormers leads to bacterial and parasite resistance. Confinement housing can lead to higher disease and parasite loads. Drugs and chemicals that are used unnecessarily create greater expense.

The opposite of conventional husbandry is organic management. True organic livestock is difficult and costly to produce. Pasture, hay, and grain fed to organic goats must be free of antibiotics, chemicals, or hormones. The same is true of bedding. Organic livestock feed is often more expensive than other feeds. The use of chemicals, wormers, antibiotics, and similar drugs is eschewed in favor of herbal remedies and strong genetics. Consequently, when life-threatening illness strikes, the organic farmer has fewer tools to treat the illness.

On the plus side, organically raised goats may be healthier and more resistant to disease through natural

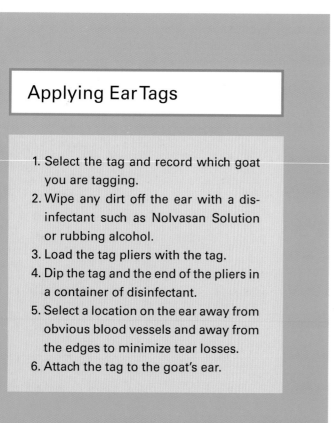

Applying Ear Tags

1. Select the tag and record which goat you are tagging.
2. Wipe any dirt off the ear with a disinfectant such as Nolvasan Solution or rubbing alcohol.
3. Load the tag pliers with the tag.
4. Dip the tag and the end of the pliers in a container of disinfectant.
5. Select a location on the ear away from obvious blood vessels and away from the edges to minimize tear losses.
6. Attach the tag to the goat's ear.

Tattooing

Supplies

Tattoo ink

Tattoo pliers and characters

Restraining device

Toothbrush

Baking soda

Alcohol pads

1. Load the tattoo letters/numbers into the pliers. Check that they are correct by punching a piece of paper.
2. Have a second pair of pliers loaded for the tattoo on the other side.
3. Place smaller animals on your lap, between your knees, or in a disbudding box. Having an assistant can be very helpful. Place older animals in a stanchion.
4. Clean the ear or tail completely, using alcohol. Clip the ears if desired.
5. Apply ink to an area slightly larger than the tattoo on a flat section of the ear or tail between the ribs of cartilage.
6. Position the tongs. The bottom of the characters should be aligned facing the bottom of the ear or the outside edge of the tail.
7. Squeeze firmly and release. Be sure to lift the tongs out straight so that the tattoo doesn't get distorted.
8. Reapply the ink and rub in firmly with an old toothbrush.
9. While not necessary, rubbing baking soda onto the tattoo can improve the set of the ink.
10. Sanitize the tongs and characters between goats by dipping in alcohol or disinfectant.

selection. Although organic goat products typically cost more to produce, they also tend to command higher prices. Goats on strict organic diets provide food that a family can trust.

Sustainable agriculture tries to combine techniques from both organic and conventional systems. Efforts are made to reduce the amounts of antibiotics and chemicals in feed and treatments. However, both antibiotics and chemicals may be employed when necessary. The sustainable goat farm tests for parasites and treats only when worm loads indicate a need. Feeds do not normally contain drug additives but may not contain certified organic ingredients. Goats are put out to pasture whenever possible.

FEEDLOT MANAGEMENT

A feedlot is an operation where animals are raised in such numbers that manure accumulates and prohibits significant groundcover. Municipalities have regulations regarding feedlot management. Manure accumulation and bare spots next to the barn to the contrary, most small family goat farms do not constitute a feedlot.

Check with your state Department of Agriculture if you believe your operation may need a permit. The main concern with concentrated animal operations is the potential for surface-water and ground-water contamination. Other public concerns include odor, traffic, and aesthetics. Regulations focus on the number of animals on the site and the methods in place for handling waste products.

While most goat operations are exempt from feedlot rules, it makes sense to voluntarily follow the best practices in your operation. Running water and drainage should be away from the animal housing and yards. Divert runoff so it does not cross these areas. Be aware of any high concentrations of manure that may leach runoff into bodies of water. A buffer strip of land between the lot and any stream or pond reduces the possibility of contamination.

Caprine management systems range from conventional at one end to organic at the other. A sustainable management system falls somewhere in between. *Shutterstock (above), Jen Brown, Terrapin Acres (opposite)*

DEAD ANIMAL DISPOSAL

An unfortunate aspect of raising livestock involves dealing with death. Depending on the size and scope of your operation, mortality may be a rare experience or a regular chore. During kidding season, owners lose kids through premature birth, accident, or illness. Breeding season is hard on bucks, and kidding takes its toll on does.

Municipalities have laws regulating carcass disposal. Rules take into consideration public health and safety, including air or water pollution, spread of disease,

nuisance odors, and pest attraction. Check with the state Department of Health to learn approved disposal methods. State law typically prohibits throwing even small carcasses into home garbage pickup.

Burial is common when carcass numbers are small and where climate permits. The carcass should be at least 3 to 6 feet under the surface to discourage scavengers. It should also be at least 5 feet above the water table to prevent ground water or well contamination.

Cremation, or incineration, is usually regulated. Open burning is prohibited in most locations. Incin-

eration can be accomplished any time of year and has a low biosecurity risk. Disadvantages include the cost of equipment and the smells associated with burning flesh. Consider your neighbors.

Biosecurity risks have decreased the popularity of sending dead livestock to rendering and pet-food plants or fur farms. Moving carcasses from one farm to another poses risks to both properties. There is no way of knowing what type of illness killed the animals previously transported by the vehicle coming to pick up your dead animals. Many disposal firms won't take

goats or sheep for fear of scrapie. Nevertheless, it is worth checking out local fur farms, wild game refuges, or pet-food plants in your area. Make these contacts before they are needed.

Composting as a method of disposal is gaining in popularity. Environmentally friendly, inexpensive, and low-risk in terms of biosecurity, composting should be considered whenever possible. Layer at least 1 foot of sawdust, bedding, or another carbon source below and around the carcass. Properly done, composted carcasses give off few odors and present very little biohazard.

CHAPTER 4

GETTING YOUR GOATS

*There is no house possessing a goat
but a blessing abideth wherein.*
—Muslim saying

Before you purchase a goat, it is important to know why you want it and what you are going to do with it. Consider the following questions:

- What do I want from my goats—meat, milk, fiber, or companionship?
- How many animals do I want to take care of?
- What space—both housing and pasture—do I have?
- What goat breeds are available in my area?

If you are expecting to have milk, you definitely don't want a wether (a castrated male). If you don't want to milk twice a day for 305 days a year, that wether may be just what you need. Some breeds, such as the Nubian, are good meat and milk animals. Like dairy cows, most dairy goats can be used for meat as well as milking purposes. Boer goats, traditionally a heavily muscled meat

breed, can produce rich milk in quantities sufficient for a family. There are people who milk Pygmies, traditionally a novelty pet, but most choose the more dairy-oriented Nigerian Dwarf if they want a miniature goat to supply family milk. Rare breeds allow you to help preserve unique bloodlines.

EVERYONE IS AN EXPERT

All goat owners have a favorite breed. Don't let the preferences of one breeder determine what is right for you. Visit fairs and shows. Join a goat club, if only for the newsletter. Go to a local conference. Read books and journals.

Opposite and above: Consider what you want from your goats before you select your breeds. *Jen Brown*

Even with prior planning, your interests may change as you get to know individual goats. I started out with a definite preference for the attractive, long-eared Nubian. I also had the idea that I would never own a tiny-eared LaMancha. Yet today, sweet, inquisitive LaManchas far outnumber Nubians in my barn.

One Internet source that has been a godsend is the GOATS list out of Washington State University (see Appendix). A wide variety of knowledgeable goat keepers contribute to this list. They are helpful to both beginners and advanced goat keepers, but beware: You rarely get a firm consensus of opinion.

A fellow goat owner told me, "Just when I find the reference for how to handle a certain problem, the goats show me that they haven't read the book!" When I think I can learn no more about goats, it will be time for me to quit raising them.

AUCTIONS

Once you've settled on a goat breed, your next step is to research your options as a buyer. Auctions are one of several possible sources. The auction barn is a decent place to sell goats but rarely a good place to buy one. This is where breeders sell their cull goats—inferior animals

Don't judge based on looks alone. The LaMancha, with its strange-looking ears, has a loveable personality. *Jen Brown*

that have been removed from the herd. Most of these animals end up as meat. Some nice goats do show up at these sales; usually, they are owned by someone who isn't a breeder and doesn't want to go through the trouble of marketing the animals.

If you do look for goats at an auction, examine the animals before they reach the sale ring. Run your hands over each goat and perform a thorough investigation. A healthy animal is lively and has a shiny coat and clear eyes. Watch for limping, swollen joints, or a misshapen udder. Look at lumps carefully to determine if they are infected abscesses or simply vaccination lumps. Manure from a healthy goat is firm and pelleted. Extreme skittishness may indicate poor temperament.

Those cute kids or that lovely-looking goat could have hidden defects. Cull goats may have extra teats, the inability to breed, or chronic illness. Don't be persuaded

by a low asking price. If the goat is too thin, has a rough coat, or has obvious sores or abscesses, it may not be cheap in the long run. You could pay down the line in veterinary expenses and heartache.

On the other hand, a specialty goat auction, common with meat and fiber goats, is typically a good place to both buy and sell stock. These auctions provide pedigrees, papers, and owner information. Some goat organizations hold auctions for fundraising or breed promotion.

BREEDERS

Most breeders sell not only the goat but also their knowledge. Some good places to meet breeders are clubs, shows, and exhibitions. To talk to breeders at a fair, the best time to approach them is when the animals are *not* showing. Shows can get hectic. The owners are frequently busy and lack the time to answer basic questions. If they are busy, ask for a business card or a time you can come back to talk.

Once you have the breeder's full attention, ask away. What diseases have been problems in the herd? Have the goats been vaccinated or wormed? What diet and feeding schedule have the goats been on? The answers will guide your management of any animal you buy.

PRIVATE PARTIES

Private owners who are getting out of raising goats, or those who keep just a few for family milk and are selling the excess, may have nice animals for sale. These goats are generally better socialized than those from a large commercial herd. You may not receive as much information about bloodlines, but if you don't plan to go into the breeding business yourself, there are real bargains to be found through the newspaper or online advertising.

COSTS

"In point of fact, there is no correct and reasonable price for goats. The price of a goat is whatever two parties agree to, then and there; nothing else counts. Those thinking otherwise are both gullible and vulnerable; accordingly, they are certain to suffer the consequences of their shortcomings, usually sooner than later."

—Frank Pinkerton

Beware of cheap kids. They may be cute but they may not be so affordable in the long run. *Jen Brown*

Goat prices vary as much as the animals themselves. Unpapered pet goats are usually the cheapest. Like other things in life, you get what you pay for. Research the prices in your area the same way you researched the breed you want. Breeders put ads in goat publications and online; many times they have complete sales lists of their available goats. Read the newspaper classified ads, keep an eye on sale barn reports, and go to auctions as an observer to give yourself an idea of the local market.

If you see an ad for five-dollar newborn bucklings, remember that to make good pets, they will require bottle feeding, neutering, and disbudding. These animals are being sold because they are excess from a large dairy and do not suit a novice pet buyer's needs.

Registration is an important factor in pricing. Some goat registries are breed specific. Some promote the commercial interests of the breeder; others concentrate their efforts on showing. Some just provide identification and documentation of ownership. Registered goat prices therefore have a wide range. The average is $200 to $400 for registered doelings; high quality papered bucks are often more expensive. Rare breeds and show animals, particularly champions and their offspring, can cost $1,000 or more. That said, registered stock isn't a sure bet. Genetics are based on percentages. Be aware that hidden genes in the parents can manifest in the offspring. Some breeds and lines reproduce consistently. Other breeds just don't reproduce as true to their phenotype.

Unregistered stock can be as nice as—or nicer than—goats that have papers. Remember the axiom "What you see is what you get." Make sure you like the conformation and looks of any goat you are buying. If possible, check out the goat's parents or siblings.

Numerous other factors can influence the price of a goat. Some breeders follow a protocol to prevent a devastating disease known as caprine arthritis encephalitis (CAE). Kids raised using these prevention techniques are more expensive than kids raised with traditional methods because of the extra work involved. The added value this method brings often offsets the additional expense.

HANDLING THE TRANSACTION

Once you decide to purchase a goat, how the transaction is handled will affect your satisfaction with the goat and the deal. Follow simple rules of courtesy.

If you are buying at an auction or from a person giving up a pet, many of these precautions are unnecessary. However, good breeders offer more information with their goats and generally charge accordingly. These tips will help you get the most out of your purchase:

- **Be clear in your expectations.** Talk to the breeder. Clarify your expectations about temperament, health status, body conformation, and milk production.

- **Settle health expectations before the sale.** Educate yourself about the conditions affecting these goats and decide your comfort level. If you want livestock from a disease-free herd, ask about herd health status. If you require blood testing, be prepared to pay the owner up front. Testing can be expensive, but if you feel strongly about diseases, consider the money well spent if the goat shows a positive result and you must reject it.

- **Decide your purchase limits before seeing the goat.** Goats are loveable. The excitement of getting a new one has clouded many an experienced eye. Clarify your expectations before seeing the live animal. Tell the seller if you are only willing to spend a certain amount. If the price is higher than you can afford, politely thank the breeder and look elsewhere.

- **Follow up on your inquiries.** It is easy to send an email to a farmer asking about goats for sale. Some breeders put a lot of time and effort into their reply. It is only polite—and will give you an advantage if you wish to deal with this breeder again—to express your gratitude for the information. It is appropriate to say, "I guess that isn't what I am looking for right now." Then the seller won't hold back the sale to someone else because you seemed interested and asked first.

- **Make an informed decision based on personal observation.** The sale is not complete until you say it is. A farm visit allows you to observe the health of the goat and its herd mates. Ask to see the sire and dam, if available. Check the goat for lumps, bumps, parasites, and general physical condition.

- **Negotiating is fine, but don't insult the seller.** At the farm, discuss your sale based upon direct observation. If there is something you do not like about the goat, tactfully explain your concern to the seller. It is then okay to offer a lower price if you are still interested. It is also okay to say "thank you for your time" and walk away.

- **Pay promptly.** If the seller has asked for a deposit, send the funds as promised. Payment in full is expected before the animal is shipped or before you leave the farm, unless a payment plan is prearranged.

- **Deposits are nonrefundable.** Do not expect your deposit money to be returned if you back out of the sale. The breeder may have turned down other buyers and has cared for the goat while holding her for you. (Deposits are usually returned if the goat is unavailable due to kids not being born or the sudden illness or death of the animal.)

- **Be on time.** Set farm visits at a convenient time for you and arrive when you say you will. Ask if the seller has any time constraints if you hope to visit longer than the time it takes to pay for and load the animal.

- **Pick up the goat in a timely manner.** Many farmers have taken a deposit on a goat only to find they are still feeding and housing the animal six months later. If you can't take the goat, contact the breeder to explain the problem. Negotiate boarding if you still want the animal, or release the seller from the deal. If kids have been born to a doe while she stayed in the care of the seller, those kids will be left with the seller unless previous arrangements were made.

- **Communicate!** Once the goat has reached your farm, whether having been shipped by the seller or picked up by you, communicate with the seller. Let the breeder know the animal arrived safely or if it is showing signs of shipping stress such as diarrhea or respiratory problems. Follow any treatment recommendations. If the animal is unacceptable for any reason, tell the breeder immediately so you can seek a remedy.

When buying a goat, take care to handle the transaction smoothly. *Shutterstock*

TO REGISTER OR NOT TO REGISTER

If you purchase goats that are qualified for registration but aren't yet registered, you'll have to decide whether to register them. Consider your breed, the reason you are keeping goats, and what you will do with the offspring.

Registration papers are usually required if you want to show your goat. They're also useful if you intend to eventually sell the animal or its offspring. Potential buyers appreciate being able to trace parentage and see how an animal fits into its particular breed. If you are a breeder, keeping up with registrations will enable you to attract a wider range of buyers.

CHAPTER 5

HOUSING YOUR GOATS

If a fence will hold water, it will hold a goat.
—Proverb

Bucks can be very hard on fences—especially during rut.
Jen Brown

Goats are tremendous escape artists. A solid fence is your best defense against an inquisitive, foraging goat. Avoid improper bracing of the fence with too few posts or placing the posts in soft ground. Bucks will lean on a loose fence panel until it is low enough to walk right over, and such adventures can result in out-of-season kids and damaged shrubbery.

Woven wire and chain link are also good options for fencing material. A fence gets more of a workout when located near greenery desirable to the goat. For this reason, these fences are best for larger areas or for younger goats that aren't heavy enough to damage the fence.

Light-duty fences with a single "hot" electric wire running alongside discourage leaning and escape. Otherwise, the recommendation for electric fence for goats is seven strands. Goats may be trained to respond to fewer wires. A modified New Zealand fence of four strands can work.

One electric fence on the market features woven wire. I have heard secondhand stories of sheep or goats getting tangled in the wire and shocked repeatedly, although many breeders report that it works well for their animals.

When first shocked by an electric fence, a goat frequently charges through it rather than backing away. Goats may be trained to respect the fence by starting them in a small, temporary enclosure surrounded by electric wire. Once the goat has experienced a shock, it quickly learns to avoid it. The trained goats are then ready to go out into a larger electrified area.

Cattle panel gates can be made stronger by twining two metal pipes or chain-link fence rails through the wire. *Jen Brown*

Panel fences come in three main sizes. Hog panels are 3 feet high and have a narrower spacing at the bottom than the top. These fences are nice for kid pens since the kids are too small to climb over them, and the owner can easily reach into the pen to work with the animals.

Cattle panels are my favorite for small yards. They are 4 feet high and hard for adolescent or adult goats to jump. The spacing is a uniform 6 inches except for the lowest two levels, which are 4 inches. The wider spacing can be a problem because small kids can wiggle through the fence.

The most expensive panel—and the best for general use—is the combo panel. This variety has the 4-foot height of the cattle panel and the dense bottom spacing of the hog panel.

Depending on the number of goats testing the fence and how badly any particular goat wants to escape, gate security can be a real chore. I'm still trying to find the

Hardware clips make for inexpensive latches. *Jen Brown*

Twine is always available around the farm for quick repairs or temporary pen ties. *Jen Brown*

Heavy gauge wire is a longer lasting alternative to twine and less costly than commercial fence clips for tying fencing of many types to t-posts. *Jen Brown*

ideal gate clip that allows me to access the pen easily. One-hand operation is optimal. I usually use hardware clips, which are strong enough to handle fence pressure and don't break too frequently.

Our state fair goat pens used to have latches that some goats opened with ease. One champion LaMancha had to be put in a specially secured pen at the fair since she routinely took herself and her pen mates for walks down the center aisle after opening the latch!

Bungee cords work for short-term latches. The goats like to use their mouths on them, but they hold for limited periods. On the downside, the elastic part of the cord ages quickly, and certain goats enjoy chewing them apart. In a pinch, baling twine is always cheap and easily replaceable. Just remember, some goats like to work the knots out with their agile mouths!

TIPS FOR A GOOD ELECTRIC FENCE

1. **Attach staples carefully**. A staple hammered through insulated fencing will cause a hidden short.
2. **Avoid bottom wires grounding out**. Wires in contact with wet vegetation or snow greatly reduce the charge. Allow for bottom wires to be shut down in wet snow or grass.
3. **Beware of old fences**. Running your electric fence along an existing fence line is tempting; however, the old wire has a way of contacting and shorting out the new electric wire.
4. **Face the sun**. If using a solar panel charger, face the charger toward the sun. Inadequate sunlight means a poorly charged fence.

Tension for electric fencing is crucial. These are tensions bars for a gate on a high tensile or "New Zealand" type of electric fence. *Terrapin Acres*

One or two electric wires may be offset from a non-electric fence to improve the barrier. *Terrapin Acres*

5. **Use good insulators.** Plastic and sunlight don't always mix well. Poor quality plastic will turn white or clear with extended exposure, while good insulators are treated to avoid breaking down.

6. **Keep wires apart.** Allow at least 5 inches between wires and be sure that any spacers or strainers are well separated and cannot cross each other.

7. **Don't mix metals.** Combining copper and steel weakens the fence as electrolysis corrodes the wires.

8. **Provide proper grounding.** Good grounding requires several 6- to 8-foot well-attached galvanized rods and a complete circuit from fence to ground.

9. **Repair damage promptly.** Kinked or flattened wire breaks easily. Damaged sections of fence should be spliced with a fence splicer or a hand-tied square knot.

10. **Space posts and ties properly.** Electric fencing should be elastic enough to spring back when hit by animals. If the electric wire is too tight or the fence is built with too many posts, the fence, insulators, or posts can break.

11. **Think bigger.** If you are tempted to skimp, think again. A charger with no "zap" won't impress the livestock, and a thin wire will carry a thin charge. Even with a large charger and heavier wire, the electric fence is still the least expensive large-area perimeter.

12. **Use a voltmeter.** Check the fence with a voltmeter (instead of your hand).

SHELTER

When protected from wind and wet, goats can withstand extreme temperatures. The most basic shelter that meets these criteria works for goats. Any variations are for the comfort and pleasure of goat owners. My goats have lived in sheds, calf huts, a barn with dirt floors, a barn with concrete floors, a converted grain bin, and dog houses. Imaginative owners have housed their animals in culverts, travel trailers, chicken coops, and wherever the darn things decided to make a bed!

The recommended living space is 16 square feet per full-size goat. More crowded conditions are feasible. Goats like each other's company and congregate by inclination. Allowing more space, however, spreads out their waste and helps keep the bedding cleaner. Crowding also increases the risk of disease, especially parasites.

Goats and other small ruminants are generally hardy animals that can withstand very cold temperatures, so long as they also have shelter from wind and wet. *Terrapin Acres*

Some goats have humble abodes *(top);* **others share large, open barns** *(above). Jen Brown*

Calf huts are easy to clean, moveable for cleanup, and relatively inexpensive. Drawbacks to the least expensive models include cracking and quick breakdown of the polymer plastic or fiberglass from sun exposure. Huts can be difficult for owners to maneuver within when catching their goats.

Barns and sheds are advantageous. The accessibility of a pen in a barn beats the hut hands down, especially when the goats are kidding or sick. My own barn has a concrete floor that was there when we bought the farm. This material is hard to clean and requires a deep bed of straw or other bedding for comfort. A dirt floor is better for absorbency and comfort for the animal's feet.

Poplar Hill Dairy is particularly well designed. The animal pens have dirt floors, while the aisles for human traffic are concrete. The concrete level is higher than the pen level, which allows bedding to build up in the pens without spilling into the aisles. Along the aisles are fences that double as feeders. Hay can be placed right on the concrete for the goats to reach. Waste hay and dust can be swept into the pens for easy cleanup. Doors on the ends of the dirt pens allow access for cleaning up manure.

These bucks live in a converted grain bin. *Jen Brown, Terrapin Acres*

Calf huts have the advantage of coming in a variety of styles and sizes while being readily portable. Unfortunately, weather, animal activity, and time all contribute to breakage. *Carol Amundson, Terrapin Acres*

Goats are playful and will explore any new props you put in their yard. *Jen Brown*

FLIES

Where there is livestock, there are flies. Under optimal conditions, fly eggs mature into adults in as little as three days. Your farm may experience as many as ten to twelve generations of flies in one summer. Flies make the animals and caretakers uncomfortable—and they spread disease. In short, they are a menace!

Do not give the flies a place to breed. Keep your facilities clean and dry. The ammonia in animal waste is particularly attractive to flies. Spreading agricultural lime on floors and in bedding discourages flies. Spiders, bats, and barn swallows should be welcomed, as they help control the fly population.

Fly traps may be purchased or homemade. Bags or jugs of water containing an attractant are placed as lures. Place them where they will not be knocked over or grabbed by the goats. Some attractants, such as sugar, leave a sticky mess when spilled, and pheromone traps often smell putrid.

Sticky tape strips or streamers may be hung in the barn. They trap large numbers of flies but also catch beneficial insects. Hang fly tape away from the flight path of birds and bats.

A useful homemade fly trap can be made with an old milk jug filled with water and a fly attractant, and then capped with a trap lid. Flies enter and eventually drown in the liquid. Chickens make more attractive fly catchers although their droppings can be a nuisance. *Carol Amundson, Terrapin Acres*

I prefer to use natural fly control methods. However, hiring a pest control company is easy and certainly keeps flies at bay. You can also use your own chemical application. Chemical risks include contamination of food products, destruction of beneficial insects and wildlife, and development of resistance in the flies you are trying to control.

FEEDING YOUR GOATS

The goat ate things. He ate cans and he ate canes. He ate pans and he ate panes. He even ate capes and caps.
—Siegfried Engelmann and Elaine C. Bruner,
The Pet Goat

Myths aside, goats are picky eaters. They may taste many things but frequently do not eat what they examine. Water and hay contaminated by feces remain untouched unless the goat is driven by desperation. Some goats are suspicious of new foods and must be almost starving before they will try unfamiliar products. Other goats will reject the offering after tasting or feeling the texture. Proper feeding ensures that goats live longer, are more productive, and have fewer health issues overall. The largest annual operating expense for livestock producers is normally feed-related.

The feeding of ruminants is a science in itself. New feeds or a rapid increase of protein or fat can throw off the balance of the delicate chemical processes in the rumen and lead to potentially fatal conditions such as diarrhea, bloat, or enterotoxemia. Changes in diet should be administered slowly and monitored for signs of trouble. The basic elements of a goat's diet include water, hay, grain (also known as concentrates), and supplements.

WATER

Water makes up more than 60 percent of a goat's soft tissues. Goat milk is 87 percent water. All of this fluid needs to be replaced daily. When they are under stress, especially when traveling to a show, goats sometimes refuse to drink water. Try adding vinegar, electrolyte solution, or even Kool-Aid to entice them to drink and stay hydrated.

Have clean water available to the goats at all times. Place water buckets and troughs where the animals can drink easily but also where they cannot foul the water with waste, such as on the other side of the fence. The parasites, bacteria, and viruses that live in contaminated water can lead to lower milk production and sickness.

In a kidding pen, water buckets should be placed in a spot where there is no danger of kids being born into the bucket. Newborns as well as lively older kids may fall into a bucket and not be able to get out. Drowning is a tragic loss of a healthy kid.

Automatic waterers can be installed. A number of commercial models work with either troughs or floats and allow animals a continuous supply of water. Another option is the Lixit or nipple valve. This simple device is the same as the nipple on a water bottle used in a small-animal cage. The larger version used for goats is mounted in a plumbing line or a container beside the pen.

When temperatures drop below freezing, watering becomes a major challenge! In Minnesota, for example, temperatures stay below freezing for weeks. Goats eat snow and lick ice but not enough to meet their water needs. Heating or insulating the water container is necessary.

Heated water systems are the most common solution for winter watering. Floating heaters are placed in tanks or buckets. Non-metal containers melt or burn when contacted by the metal heating element in these types of heaters, so keep the water level high and use heaters with guards. Keep electric cords away from curious and playful goats.

Keep clean water available to the goats at all times. Water needs to be changed regularly. If you keep other livestock, particularly waterfowl, they will also use the goats' water. One ounce of bleach per fifteen gallons of water can be added to reduce the growth of microorganisms. Careful! Too much bleach can upset the rumen organisms. *Jen Brown*

Buckets with a heating element built into their structure are a great investment. The cord is hidden within the walls of the container. The water is easily accessible to the goats, as there are no floating heaters to get in their way. Larger automatic watering systems are available in heated models with thermostats that regulate heat during freezing temperatures.

HAY

Good-quality hay is important for caprine nutrition. The average goat eats 4.5 pounds of hay a day per 100 pounds of body weight. Hay provides protein and vitamins, as well as fiber to keep the rumen working properly. Fiber provides energy, particularly in milkers, young goats, and does in late pregnancy. The younger and fresher the hay, the better the quality. Good hay is expensive, but the better the hay, the

Green, leafy hay is high in Vitamin A, one of the vitamins necessary for healthy goats. They also get this nutrient from corn or good pasture. *Carol Amundson, Terrapin Acres*

less grain and supplemental feeding you'll need. High-quality hay is especially important in the dairy barn. Good quality hay can make immediate improvements in the amount and quality of milk produced.

FINDING HAY

If you don't grow sufficient forage on your property to feed your goats, you'll need to purchase it from a reliable source. For the pet goat owner, a few bales at a time from the local feed store or a neighboring farm are enough. Large operations may have hay shipped to the farm by the ton. Different parts of the country have different types of hay. In the Midwest, alfalfa, grass, and clover are common.

Stored hay loses nutrient value. A field dotted with large round bales of hay may look beautifully pastoral, but this method of storage causes drying and loss of digestibility. A round hay bale stored outside may lose 15 to 25 percent of its feed value within the year. It is better to store hay in the protection of a barn. Depending on the moisture content, hay stored under cover can retain most of its nutritional value for up to a year. Putting up

hay that is too wet leads to mold and microorganisms, which consume nutrients and release toxins that can cause serious illness. Never feed your goats moldy hay! In a worst-case scenario, overly wet hay stored in a building can spontaneously combust. The resulting barn fire would devastate any operation.

PASTURE OR RANGE

Pasture and browse can supplement hay. Don't try to maintain a goat herd strictly on pasture. Young pasture that is rapidly growing may contain too much moisture for the goats to eat enough to meet their protein and nutrient needs. Drought conditions change the composition of plants and can also concentrate toxins. Allowing pasture goats to have hay at least once a day ensures that they are properly fed.

A goat's natural eating style is to wander about and pick and choose from a variety of browse. Goats are useful for cleaning unwanted brush. Brush has lots of fiber, plus vitamins. Be aware that even the brushiest area will not survive intensive grazing for more than two seasons.

Stored hay loses nutrient value. The first things to drop are Vitamins A, D, and E. *Jen Brown*

During the summer, goats that are not growing rapidly or milking heavily may get all their nutrition from good pasture. Pregnant goats, does in heavy milk production, and adolescents usually need supplements of other high-energy feed. *Jen Brown*

Branches placed in the goat pen quickly get stripped of leaves and bark. *Jen Brown*

Tree branches are a treat for goats that don't have browse in their pasture. *Jen Brown*

SILAGE

Special warning should be made about silage, which is not recommended for feeding goats. The risk of mold, bacteria, or even excess nitrates is high. Goats are particularly susceptible to mold and other spoilage. Goat owners have learned this lesson after tragically losing a large portion of their herd to illness and death following the feeding of seemingly good quality silage.

HAY FEEDERS

Hay feeders create unique puzzles. The ideal feeder holds hay up for goats to reach comfortably while preventing scraps from falling out of the feeder and being wasted. The perfect feeder also allows goats access to hay without trapping their head or horns. The feeder is tight enough to keep smaller goats from climbing in and spoiling the hay with their dirty hooves or waste. The feeder capacity is large enough to hold a half-day's worth of fodder. Naturally, this mythical feeder does not exist. You'll need to find the compromise that works best in your situation.

Bag and Net Feeders

Typically used for feeding horses, a bag or net feeder hangs in the pen either on the fence or wall or suspended from the ceiling. The feeder must be close enough to the ground to allow the goat to get at the hay but high enough that the goat does not get tangled in the net. I have read about goat owners who swear by this feeding method. I have never used it, as this design seems to create the risk of legs being trapped and then broken by the goat's ensuing struggles. Worse still, the goat's head could be caught, causing it to strangle to death.

Hanging Feeders

One type of hanging feeder is a simple, inexpensive feeder that slips onto a fence. Portable and easy to handle, they hold one or two flakes of hay from a small, square bale. Some of these feeders are flimsy and do not hold up well to the rough treatment that goats can dish out. The wires bend and break off. Once this occurs, the feeder becomes useless because it wastes more forage than it feeds. If you just have a couple of pet goats, though, a hanging feeder is feasible. It is also a good

Bag-style horse feeders serve as temporary feeders at a fair for some young goats. *Carol Amundson*

option for a temporary isolation pen in your barn or a pen at the fair.

Fence-line Feeders

A fence-line feeder uses the fence as a barrier while providing feed at the same time. Cattle panel fencing allows goats to stick the muzzle or head through the wire to get at the hay on the other side. For larger goats or those with horns, some of the holes should be enlarged so the goats may maneuver in and out of the feeder. Be aware, however, that every enlarged hole becomes a possible escape route for kids and smaller goats.

A fence-line feeder is a common and easy hay setup.
Terrapin Acres

For adjacent pens, feeders can be used that hang on both sides of the fence. Some of these are only good as temporary feeders because they get easily banged around and battered by the goats. *Carol Amundson*

The wire in this feeder is sturdy and will hold up to hard use. *Carol Amundson*

Using fence panel feeders can be a danger to horned goats, who get their horns stuck in the fence. *Terrapin Acres*

A sneaky Nigerian kid slips through the fence. With any feeding setup, you need to figure out how to give larger goats access to the food while preventing kids and smaller goats from slipping through and fouling the food. *Terrapin Acres*

A bolt cutter is indispensable. Keep it near the barn for easy access when a goat is stuck in the feeder or fence. *Jen Brown*

Above and opposite, bottom right: Homemade feeders from wood or a combination of wood and fencing are inexpensive and useful and limited only by the breeder's imagination. *Jen Brown, Poplar Hill*

Mangers and Keyhole Feeders

A manger is a wooden trough or open box that allows goats easy access to hay without much waste. A keyhole feeder is a variation of the manger with small holes that individual goats can fit the head through to eat. Featured in many beginner goat books, this design has both pros and cons. One negative is that a goat cannot see when her head is in the keyhole. This prevents the animal from noticing a more aggressive goat coming from behind to slam her away from the food. A goat also has difficulty getting her head out of the keyhole and can injure herself if she tries to exit quickly.

Copper Needs in Goats

A generic sheep-and-goat feed used to be recommended for both animals, but that has changed within the past ten years. Studies have shown that sheep and goats have different copper needs. Sheep are sensitive to copper, and too much in their diet is harmful, whereas goats require more copper in their diet for proper immune function. Copper deficiency can lead to a whole host of health issues, including lameness, weight loss, diarrhea, poor milk production, pneumonia, abortions, and birth defects.

By the same token, it was once believed that the high copper content in horse feed was toxic to goats. It is now known that the copper needs of horses and goats are similar, and horse feed is acceptable for goats. (A word of caution: Horse feed is also high in fat, and too much can lead to obesity or urinary calculi in goats.)

Most pastures and hay types, particularly alfalfa, do not provide enough copper to meet the needs of goats. That said, do not automatically give your goats a copper supplement! Too much of any mineral can be toxic, causing health problems or even death. The best way to determine how much copper your goats are getting is by testing liver samples from dead animals. If your goats are copper deficient, give them a yearly supplement in the form of a bolus. Mineral supplements should be administered in the fall, two to four weeks before breeding.

GRAIN

Every goat needs some grain during its life. The amounts and types of grain that should be fed depend on the type of goat and the quality of the available forage. Kids and growing goats, working goats, lactating does, and those in late pregnancy should all receive grain regularly. Many pet goats and non-lactating does receive too much grain and become fat as a result. Goats generally eat as much grain as they can get. Changes in the type or amount of grain should be made gradually to avoid illness or even death.

Feed recipes abound. Commercial mixes may be purchased. Individual feed mills may sell their own mix. Some breeders feed goats a 16-percent "sweet mix" made for cattle and horses.

In a slant bar feeder, even goats with big horns can eat, yet be separated from their herd mates to prevent fighting. Plan on at least one foot of feeder space per goat. *Carol Amundson*

Goats will step in their feed when they can—causing contamination from feces and possibly spreading intestinal diseases. Keep feeders off the ground. *Carol Amundson*

GRAIN FEEDERS

There is a wide variety of grain feeders on the market. When deciding on a feeder, the goat keeper must again weigh convenience, cost, and the goats' tendency to climb and play. Consider the amount of space needed for each goat to reach the grain at feeding time. More aggressive animals often prevent meeker herd mates from eating. Owners should separate feeders and allow enough head space for all the animals to have a fair chance.

Creating a fence-line feeder can be as simple as dropping the grain on the ground or concrete outside the fence and allowing goats to eat it by sticking their heads through. This can work in an inside pen that is protected from weather. On the other hand, uneaten grain can create quite a mess outside on dirt or grass. It can also lead to increased parasite loads due to ground contamination with organisms.

The Milk Stand or Stanchion

Individual feeding stands are highly desirable for a small goat herd. You can control the amount of food each goat eats by locking the goats into individual stanchions or tying them by a feeder. At feeding time, portion each goat's ration and add any special wormers or supplements that an individual goat may need. This method is impractical in larger herds except dairies, where milking does are fed on the milk line.

Like people, different goats have different metabolic rates. Goats that have lower feed requirements are known as "easy keepers." An overweight milker needs additional grain to milk well; however, the extra grain aggravates her weight problem. One trick is to feed her beet pulp soaked in warm water. This fills her up and promotes milk production while not giving as many calories as concentrated feed.

SUPPLEMENTS

Supplementation is an evolving science, with even the experts constantly learning more about which vitamins and minerals are best for goats. Like hay and grain, supplementation needs vary by type of goat, season, region of the country, and quality of forage and grain being fed. Fed free-choice or added to a grain mix, these supplements come in a variety of forms.

LOOSE MINERALS

Loose minerals are best placed in free-choice feeders. General goat mineral formulations are available, as are specific formulations for meat herds and dairy herds. A mineral called Buck Power keep bucks in prime breeding condition. If goat mineral blends are not available, formulations for cattle or horses can be used.

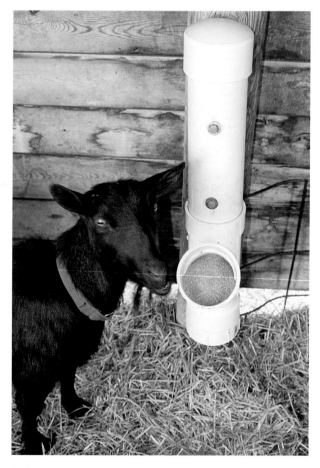

A nice version of a loose mineral feeder may be made using PVC plumbing supplies. *Barb O'Meehan*

BLOCKS

Blocks are formed of compressed quantities of salt, sulfur, cobalt, or a mix. Blocks are durable and weather resistant, although goats may cause wear and tear by playing "king of the hill" on them.

Wet blocks are feed supplements bound together in a molasses base. They are beneficial for pregnant does in late fall and winter; the closer certain goats come to their due date, the more they use their block for energy and protein. Some goats become very messy as they spend their time licking at the block and their faces become sticky with the molasses.

BAKING SODA

Baking soda deserves special mention. Giving your goats free-choice soda in their mineral feeder buffers the rumen and helps prevent stomach problems.

AMMONIUM CHLORIDE

Male goats, particularly wethers, are prone to a condition called urinary calculi in which stones form in the urinary tract from a concentration of salts. Adding ammonium chloride to the daily feed at the rate of 1 teaspoon per 150 pounds of body weight acidifies the urine and prevents stones from forming. Urinary calculi can be both painful and life-threatening, so I consider this a necessary additive.

PROBIOTICS

The addition of yeast or other probiotic mixtures can help the rumen function and improve feed utilization. For some goat owners, this is a routine addition to a feed mix. Others use it only for goats under stress or in ill health. Dry and wet versions are available commercially.

BOLUSES

Some supplements are given as a bolus, or intravenous injection, on a timed schedule to cover deficiencies in certain nutrients, including selenium and copper. Selenium supplementation, most often called Bo-Se, is recommended for goats in selenium-deficient regions of the country.

Your herd may contain goats of all ages and sizes competing for food. It is important to have a good feeding setup, unlike this one, where smaller and weaker animals can get adequate feed. *Shutterstock*

ELDER CARE

Older goats have special needs. Goats are considered aged at five years. By eight years of age, most goats exhibit signs of aging. Teeth wear down or fall out. Additional grain or liquid supplements may be required to improve overall condition. Give senior goats easy access to food and water. Arthritis or stiff joints may cause pain that prevents them from eating enough food or drinking enough water. When older goats are housed with younger animals, competition is fierce, so be sure that elders get what they need to be healthy.

BREEDING YOUR GOATS

Depending on what you want out of your goat herd, extra teats may be grounds for culling. It's not as important in a meat herd but a negative trait in a dairy herd. *Jen Brown*

The more the billy goat stinks, the more the nanny goat loves him.

—Belgian proverb

Goats are prolific and hardy. Expect healthy yearlings to produce one to two kids per year and older does to produce two to three. Five to seven kids from a single dam in a year is an unusual occurrence. The Nubian, Boer, and Nigerian are known for multiple births.

Goats with obvious defects should be culled from your herd. Never sell a cull animal to an unsuspecting buyer without full disclosure. In the case of serious defects, the animal should be put down.

Some genetic faults are more tolerable than others, depending on the purpose of the animal. Dairy farmers do not want extra teats on a milker. In meat herds, extra teats may be disregarded. Show herds require strict culling for superficial traits that would not bother a pet goat owner.

It is useful to understand basic genetics before starting a breeding program. Each parent goat randomly provides one set of genes to the kid. Think of genetic inheritance like the flip of a coin. The kid inherits a pair of genes that may be alike or different. Identical genes are known as homozygous; different genes are called heterozygous.

When a kid inherits two heterozygous genes from the parents, the dominant gene shows in the kid's appearance. The appearance of an animal is known as its phenotype. Purebred animals have been bred for certain phenotypes over many generations, giving the LaMancha its tiny ears, the Pygmy its size and body structure, and the Angora its fleece. The dominance of a trait is often apparent in crossbred animals. A LaMancha usually produces offspring with smaller ears, for example, and the muscle development and color of the Boer is typically dominant.

A recessive gene is one that hides in the genetic background. Recessive traits only become visible in the phenotype when the kid inherits two recessive genes. A dam and sire may both lack a visible trait but pass that trait to their offspring through recessive genes. Colored Saanens, now known as Sables, are an example of a recessive gene unexpectedly springing up.

Improvement is the goal of the breeder. But improvement is in the eye of the beholder. One person may be looking for increased milk production, another for more mohair, and yet a third for better attachment of the udder. Some breeders intentionally cross breeds to create new types of goats or improve the vigor of their animals. Meat breeders increase growth rates and survivability through crossbreeding. Each breeding will show improvement and regression of multiple traits.

THE BUCK

There's an old saying among goat keepers: "The buck is half your herd." The buck's impact on a herd is typically greater than that of a single doe, since he may sire substantially more kids than she can produce in a lifetime. Buck selection is both a science and an art. The first rule is to breed what you like to see. Look for a buck that excels in the traits you value and shares the good traits your does already possess.

Research available bucks to ensure that they have the proper genetics to improve your herd. Do not look only at the buck; some really ugly bucks have sired fabulous kids. Similarly, some fine-looking bucks are known to "throw" extra teats or an underbite or overbite. Check the buck's dam, sisters, and kids.

Because he will invariably produce more offspring than any single doe, this Golden Guernsey buck is "half the herd." *Phyllis Clayton, Golden Guernsey Goat Society*

WHEN TO BREED

The best conception rates and the healthiest pregnancies occur when goats are in good condition. High-quality feed, worming, and proper mineral levels are important. Too much of a good thing can reduce conception rates, though, so make sure your goats do not go into the breeding season too fat. Some breeders practice flushing, which entails gradually increasing the quality and amount of feed in the weeks prior to breeding. Flushing can help the doe ovulate more eggs.

Caprine gestation lasts 150 days. Think before putting your doe in with a buck simply because she is in heat. Look at a calendar. Some goat keepers plan matings around show season or market factors. Kidding during the hottest and coldest months is hard on the farmer and the goats.

A buck in rut follows a doe. *Terrapin Acres*

Mating Systems

TYPE OF BREEDING	DEFINITION	ADVANTAGES	LIMITATIONS
Outcrossing	Breeding two animals of the same breed but with no common ancestors for the past four to six generations.	• Brings in strong dominant traits. • Creates hybrid vigor, including longevity, better growth, and improved reproduction. • Hides bad traits by keeping them recessive.	• Greater variability in offspring. • Improvement is based on selection and availability of superior genetics.
Line Breeding	Breeding two animals with a relationship in pedigree for a low level of selective inbreeding.	• Fewer risks than continued inbreeding. • Greater uniformity of type. • Helps locks in strong traits. • Increased prepotency.	• Increased chance of recessive defects. • Slow improvement of line, especially if it is mediocre.
Inbreeding	Breeding two animals that are directly related, such as mother, father, or full sibling.	• Carrier animals can be identified and culled. • Helps detect inferior genetics. • Increased homogeneity of type in offspring. • Increased prepotency.	• Creates inbreeding depression, resulting in loss of size and fitness. • Higher risks of kids with defects. • May reduce available genetics in future relationships. • Rigid culling required.

HOW TO BREED

There are several options for breeding, each with its own advantages and disadvantages. Factors to consider include the number of goats owned, proximity to the buck's residence, type of commercial operation (if any), and whether or not the breeder needs to know exact breeding dates. You may find yourself using a combination of methods.

PEN OR PASTURE BREEDING

The easiest way to breed is to put the buck and does in a pen or pasture together and let nature take its course. A buck can service more than two dozen does with a reasonable conception rate.

Forty-five days covers two breeding cycles, and most does will conceive in this time if it is breeding season. After this, as the saying goes, familiarity breeds contempt. When does and a buck are kept together for more than a couple of months, they often lose interest in each other. Separating the sexes occasionally might help. The opposite is the "buck effect," in which a newly introduced buck causes does to cycle into heat.

RADDLE HARNESS

A variation of pasture breeding is the raddle harness, which is placed on the buck and equipped with a special crayon on the chest. (Traditionally, to *raddle* something is to mark it with red ocher.) An alternative is to spread a colored paste on the buck, with no harness; the paste must be applied more frequently than a harness crayon needs to be replaced.

When the buck has a solid mount on a doe, the marker colors the doe's rump. The more mountings, the more color. The breeder changes the color of the marker every seventeen to twenty-one days until no further marks are seen on the does. Tracking the color on the doe's rump on the calendar identifies probable breeding dates. In this way, raddle marking offers a method of tracking breedings without the need for daily monitoring. This method is used most often in meat and fiber herds kept on range.

HAND BREEDING

Taking the doe to the buck and monitoring the event can be done by keeping the buck on a lead-line or by using a small pen for the buck and doe to visit for a short time. If the doe is in heat, breeding requires five to ten minutes for the buck to mount, ejaculate, and repeat the process. You'll know the breeding has occurred when the doe arches her back sharply and the buck throws his head back, then falls or staggers backwards. Once you see it in action, you will recognize the signs.

Sometimes the doe rejects the buck or is reluctant to breed. This happens most often with virgin does. The owner may hold the doe steady for the buck to mount. The buck may also be reluctant to mount. In the (fortunately rare) case of a slow buck, an alternate buck may have to be used. Sometimes teasing the doe with another buck on the other side of the fence induces a slow buck to mount her.

Taking your doe to another farm to be bred allows you to keep fewer bucks or no buck at all. Someone who raises only a few does should try to find a local breeder offering buck service. If you have registered animals, be sure to get a service memorandum so that you can register the offspring. Costs of outside breeding vary, but if you have only a few does, breeding or lease fees are usually less expensive than owning your own buck.

A doe teases a buck from the other side of the fence.
Jen Brown

An excited buck checks a doe. *Jen Brown*

Methods of Breeding

BREEDING METHOD	ADVANTAGES	DISADVANTAGES
Pen Breeding	• Easy. • Not necessary to track heats. • Good for short-cycling does. • Good for hard-to-detect heats. • Useful for off-season breeding.	• May not know exact breeding date. • Difficult to perform prenatal care on doe without knowing breeding date. • May require additional pens, depending on number of goats being bred and number of bucks used.
Raddle Harness	• Similar to pen breeding. • Visual indicator of breeding.	• Expense of harness and markers. • Attaching harness may be tricky. • Buck may remove marker. • False marks with aggressive bucks or passive does.
Hand Breeding at Home	• Know exact breeding date. • May require fewer pens.	• Need good heat detection. • Requires close monitoring of does several times a day. • More time-consuming. • Handling the buck may be difficult.
Hand Breeding off Farm	• Similar to home breeding. • Wider range of bucks to choose from. • Eliminates or reduces the cost of owning bucks.	• Similar to home breeding. • May require multiple trips. • Travel distance and time. • May be difficult to locate breeder offering service. • Breeder may require certain health tests. • Possible to bring home diseases.
Artificial Insemination	• Wider range of bucks to choose from. • Faster genetic improvement possible. • Eliminates or reduces the cost of owning bucks.	• Initial equipment expensive. • Usually done by goat owner. • Good heat detection required. • Conception rates vary. • Conception rate dependent on technique.

ARTIFICIAL INSEMINATION

To bring in new genetics, try artificial insemination (AI). You can keep frozen semen stored for many years in liquid nitrogen. AI is highly technical but readily learned by any goat owner with the interest. Unlike with cattle, there are few professional inseminators who come to the farm and breed your goats.

ESTRUS

Does typically reach puberty at about six months of age, although some become fertile sooner. Separate does from bucklings by three months of age.

Most goats are seasonal breeders. The season is influenced primarily by light, with its onset triggered by decreasing hours of daylight. In the Midwest, the breeding season starts in September or October and runs through February.

Some breeds are easier to breed out-of-season than others. Myotonics, Boers, Pygmies, and Nubians are known to breed year-round. Individual animals may be exceptions, and some breeders select their goats for out-of-season breeding.

During breeding season, does cycle into estrus every eighteen to twenty-one days. The first step in breeding is to have a doe in standing heat. This heat may last only a few hours or up to three days, depending on the animal and the time of year. Once the doe ovulates, the egg has a viable life of about ten to twelve hours.

Some does are clear about their desire to mate; others are shy or have quiet cycles. Goat keepers use several methods to detect heats in the herd. An otherwise healthy hermaphrodite goat is incapable of breeding and producing young. However, these animals are sometimes used as teaser goats to detect estrus. Because they often act buck-like, hermaphroditic goats react to does in heat just as an intact buck would—signaling to the owner when a doe is ready to go to a suitable mate. A vasectomized buck can serve the same purpose since his hormone levels remain high, unlike those of the castrated wether.

If there is no buck on the property, owners may use a "buck rag" to help detect heat. This method requires a rag, a jar with a tight lid, and a trip to visit a buck in rut. Rub the rag all over the buck's head and belly. Put the rag in the jar and cover tightly. Back home, open the jar

Sticky hair around the tail is one sign of estrus in the doe. *Jen Brown*

The smellier the buck and the better you rub his scent onto the rag, the better the lure of your "buck rag." *Jen Brown*

under your doe's nose several times a day. When the doe gets excited or tries to get into the jar, it is time to visit the buck.

In herds that need out-of-season or timed kidding, does can be brought into heat using medication available from your veterinarian.

Heat Signs in the Doe

- Allowing other does or wethers to mount her.
- Calling out or crying frequently for no reason.
- Exhibiting a drop in milk production.
- Fighting.
- Flagging (holding the tail high and frequently wagging).
- Acting more affectionate.

- Losing interest in feed.
- Mounting other does.
- Exhibiting a mucus discharge from vagina.
- Standing by the fence closest to the buck pen.
- Exhibiting a swollen or pink vulva (rear end).
- Exhibiting tail hair that is wet, sticky, or clumped together.

VISIT, LEASE, OR OWN?

Bucks expend a tremendous amount of energy breeding. Some bucks go off feed and lose weight during breeding season. Their single-mindedness can be dangerous. Most health problems with bucks occur during this time. Not keeping a buck reduces feed costs, the vet bill, and housing needs and prevents the occurrence of unwanted breedings should the buck escape his "escape-proof" enclosure. For these reasons, some owners elect to visit or lease the services of a buck rather than keeping their own.

VISITING THE BUCK

Long before breeding season, you should talk to breeders in your area who offer buck service. Take into account pedigree, cost, distance, and herd health. Some breeders offer boarding for does, but more commonly the doe owner brings the doe to the buck when she cycles.

The breeding season is hard on bucks. Fighting may lead to bloodied heads, and the energy expended mating depletes their reserves.
Jen Brown

Because heat times vary among does, watch the doe and record how long her cycle lasts.

Biosecurity has made it more difficult to find breeders offering buck service. Conditions that would prevent your doe from showing at a fair (abscesses, ringworm, lice, respiratory illness, diarrhea) should also prevent you from taking her to someone else's farm. Also, let the owner know if your doe has horns. Some breeders will not expose their buck to the risk of goring. The reverse is also true.

Have all registration papers and health certificates handy. Some breeders ask for these and check tattoos or ear tags before giving you a service memorandum. Expect to pay up front for a breeding service. Most breeders allow you to bring the doe back for another breeding at no charge during the same season if she does not become pregnant. Buck owners cannot be held responsible for does that abort, reabsorb the fetus, or do not settle unless a health inspection shows that the buck is infertile or is carrying a specific disease that caused the termination of the pregnancy. A breeding contract protects both owners from any misunderstandings.

The polite doe owner watches carefully for the first sign of heat and communicates its arrival as soon as possible to the buck owner. Actual breeding may take only minutes, but preliminary courtship and the paperwork following the act can be time consuming. Be sure to factor "visiting time" between the goat owners into your schedule as well. Showing up at the farm without warning or insisting on coming at a bad time for the breeder may cost you the opportunity to breed there again.

There are many factors to consider when deciding whether you want to own, rent, or lease a buck for your breeding program. Cost, convenience, and housing and fencing needs are just a few. *Jen Brown, Poplar Hill Dairy Farm*

LEASING A BUCK

Some breeders find it more convenient, especially if multiple does are involved, to lease the buck rather than take multiple trips to the buck farm. Buck owners may require an inspection of the leasing premises and additional health-status information. An agreement should be reached beforehand about what happens if the buck becomes ill, injured, or dies. Usually, the leaser feeds, houses, and cares for the buck. This may include covering veterinary costs if the animal becomes ill.

For pedigreed animals, goat associations have buck lease forms that can be filed with the association. The leasing farmer may then fill out registration papers and submit them without service memos. Service memorandums are required for any breeding that does not have a lease form on file at the office. Some breeders request that only does be registered out of the breedings. This arrangement should also be discussed before breeding or lease.

The older the buck, the more difficult it may be for him to relocate to a new farm and adjust to new feeding or housing arrangements. A buck that was content at home can go off feed, become ill during transport, or pine for his herd mates. I generally lease young bucks only to herds I know well or ones with only a few does.

OWNING A BUCK

If you decide to keep one or more "boys," it is important to know what to expect. The buck is very different from the doe. Even as kids, some bucks show breeding behavior. Intact male goats become increasingly "buck-like" as they age and hormones kick into effect. In certain breeds and in locations with strong seasonal effects, the buck's rutting behavior ebbs and flows with the season. In other breeds or in temperate zones, the buck may remain in rut most of the year.

Characteristics of the Buck

TRAIT OR BEHAVIOR	CHARACTERISTICS	BREEDING FUNCTION	NEGATIVE ASPECT
Aggression	• Rears up. • Butts. • Attacks fence separating him from does. • Challenges other animals. • Mounts other bucks or animals.	• Attracts does. • Establishes position in herd.	• Hazardous to caretakers and pen-mates. • Requires strong fencing. • Need to separate horned and dehorned bucks to avoid injuries.
Frequent erections	• Exposes the end of the penis. • Inserts penis into mouth.	• Asserts dominance. • Attracts doe.	• Disconcerting for some owners. • Embarrassing for visitors.
Increased verbalization	• Moans and "talks" to does. • Blubbers and wags tongue.	• Attracts doe. • Encourages doe to stand.	• Sound can carry a great distance and disturb neighbors.
Lip curling (Flehmen Reaction)	• Lifts upper lip into a grimace. • Accompanies urine "tasting".	• Exposes a sensory organ in the upper lip. • Detects estrus pheromones.	• Odd appearance.
Scent	• Gives off strong "goaty" odor. • Rubs scent on anything or anyone within reach.	• Attracts the doe. • Stimulates the doe to come into heat.	• Difficult to remove from skin and clothing. • May affect milk quality if buck runs with the does. • Offensive to some people and downwind neighbors.
Stamping and pawing	• Paws the air with front foot. • Uses foot to hit at does or the fence.	• Encourages the doe to stand for service.	• Painful when applied to humans.

TRAIT OR BEHAVIOR	CHARACTERISTICS	BREEDING FUNCTION	NEGATIVE ASPECT
Urination	• Urinates on front legs and face. • Sprays urine into mouth. • Sprays urine indiscriminately.	• Further attracts does by increasing pheromones.	• Urine scald on face and legs may leave buck's skin raw and irritated. • Can cause goat owner to change clothes (yet again) after being caught in the spray.

The smell that most people associate with goats is the smell of the buck. This smell comes from oils secreted by scent glands located mainly at the base of the horns. This scent can be very strong and offensive to some people. Angora bucks have the scent gland not at the base of their horns but in their wool, so their smell may not be as strong.

In Greek mythology, Satyrs are half-goat, half-man creatures and figures of fertility. This is no accident. The ancient Greeks knew their goats. Anyone observing bucks for any length of time sees that they are obsessed with breeding. During rut, bucks are odiferous, noisy, and aggressive. In other words, they stink, blubber, and generally act rude and obnoxious! The chemical bromine, which has a strong odor and stings the eyes and irritates nasal passages, was named after the Greek word *bromos*, meaning "the stench of he-goats."

This stench stays on anything it touches, including any parts of your body and clothing that come in contact with the buck. You will, at some point, need to remove this scent to be socially respectable. Cleansers that are supposed to work include Fast Orange Hand Cleaner, goat milk soap, Listerine, and toothpaste. Febreze in the wash is a must after working with the boys.

A buck in rut frequently urinates on his face and front legs, resulting in a condition called urine scald. *Jen Brown*

BYPASSING THE BUCK: ARTIFICIAL INSEMINATION

To render the above complexities irrelevant, consider AI as a means of breeding. Unfortunately, caprine semen processors are limited in number. Buck semen is usually collected at on-farm visits by the processor, who may visit a region only once a year. When an individual farm doesn't have enough bucks to make the stop affordable, clubs sometimes organize a "buck collection," which functions as much as a social get-together for goat owners and an opportunity to semen shop from the processor's collection.

A buck in rut courts does by wagging his tongue (*above*). Another facial move is to curl the front lip, exposing scent glands—called the Flehman Reaction (*left*). *Jen Brown*

There are no professional goat inseminators in most areas. Goat owners learn AI through classes, books, and other goat owners. Classes are taught at shows, conventions, and other gatherings of breeders. Unlike with cattle, semen is introduced to the goat through a speculum and sight rather than by feel. It is an art as much as a science, with widely variable success rates between breeders.

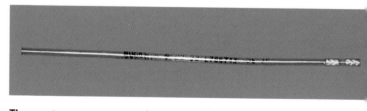

The most common storage for goat semen for artificial insemination is ½ milliliter straws. *Carol Amundson*

Have a plan in place for what you will do with the kids once they arrive. *Jen Brown*

PLANNING AHEAD

Plan ahead for the kids. As you read about raising kids, neutering, disbudding, and marketing, ask yourself the following questions:

- Will the kids be bottle-fed or raised by the dam?
- What vaccinations will I need?
- Will any buck kids be kept from this breeding?
- Will any bucklings be wethered, and how will that be done?
- Do the kids need to be disbudded, and how will that be done?
- How many kids will be kept in the herd, and how many will be sold?
- What type of markets are there for doe, buck, or wethered kids?

Buck semen freezes well and may be stored for years in liquid nitrogen tanks for artificial insemination use.
Carol Amundson

CHAPTER 8

KIDS AND KIDDING

Raising kids is part joy and part guerilla warfare.
—Ed Asner

There are no caprine pregnancy test kits on the market with the exception of costly hormone testing; however, there are other ways to determine pregnancy. The first sign of a successful breeding is that the doe will fail to come back into heat eighteen to twenty-one days after the mating. Breeders say that the doe has "settled."

Ultrasound is useful when a doe was pen bred or if there is some doubt about pregnancy. Forty-five to sixty days following breeding is the best time to check for pregnancy using this method. Some vets own a real-time ultrasound machine, the same type used for human ultrasounds. It is exciting to see the ribs and beating hearts, count kids, and even estimate time of gestation.

CARE OF THE PREGNANT DOE

At least sixty days prior to the due date, examine the doe for fitness and remove a milking doe from the production line. The final days of pregnancy are known as the "dry period." Some goats require longer dry periods, but all does need at least two months off between lactations for the kids to grow and develop. If it is part of your herd practice and your veterinarian recommends using a "dry cow" udder infusion—a solution of antibiotics to help prevent infection or cure any undetected issues—do so on the last day of milking.

Start slowly increasing the amount of grain in the doe's ration or give her better hay as her condition warrants. Some dairy farmers bring the doe onto a milk stand twice a day to accustom her to the routine if she has never milked before, to examine her for problems, and to give specific rations.

In rare cases, a doe continues to milk in spite of the best efforts of the owner to dry her off. Some goats milk through their pregnancy and do not produce colostrum, which is normally the first milk produced

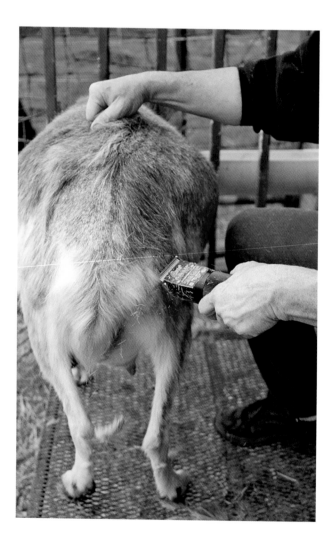

by a new mother. Colostrum is thicker than regular milk, yellowish in appearance, and full of proteins and antibodies that protect the kid until its immune system is fully functioning. In its absence, you will need to take special steps (detailed below) to ensure the kid's survival.

Thirty to forty-five days before the blessed event, give the doe annual vaccinations and shots. Ten to fourteen days before the known due date (or two weeks before the first possible date if the doe was pen bred), the doe should have a "pregnancy clip," also known as "crotching." This haircut gives the breeder an easier view of the back end of the doe to check for signs of parturition. Trimming around the tail and vagina keeps the doe cleaner at delivery. Trimming the belly and udder helps newborn kids find the teats and nurse successfully.

This is the time to move the doe to her separate kidding pen if you desire. Some goat keepers like to give the doe and offspring private quarters for delivery and bonding without interference from the herd. Other goat owners believe that the doe is less stressed when she stays in familiar surroundings. Certainly, if she is moved to a pen by herself, she should be within sight and sound of the herd to prevent loneliness.

Some farmers find a simple baby monitor useful. With the advent of reasonably priced video models, you can even see a view of the pen. Monitors are helpful for unusual situations or for detecting problems early.

SIGNS OF LABOR

There are a number of ways to tell that your doe is freshening, or approaching kidding. She may remove herself

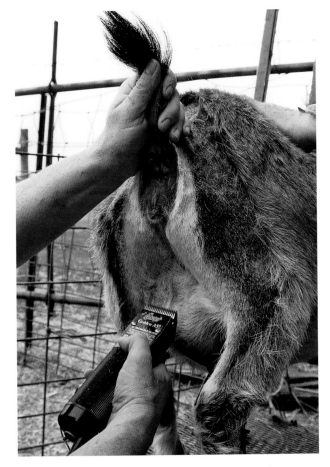

The pregnancy clip includes shaving from hips to pins *(opposite)* and shaving the udder *(above). Jen Brown*

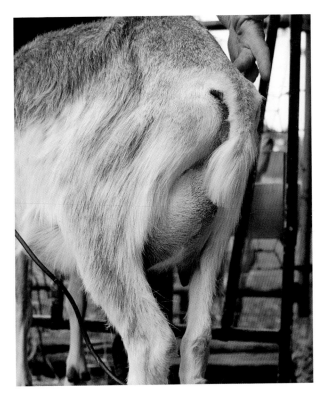

The pregnancy clip helps you detect hollowing out at the tailhead as kidding approaches. There are two ligaments beside the tail. These feel like tight bands most of the time. A sure sign that kidding is immanent is the hollowing out beside the tail. When you cannot feel the ligaments at all, expect kids. *Jen Brown*

Pregnancy Clip

from the herd. Another doe may become more friendly and anxious to have her caregiver present. She may become vocal or testy with herd mates and others. One older doe of mine would toss any passing cat or chicken across her pen during the final hours before delivery!

You can watch for bagging up, or the filling and tightening of the udder with milk, but do not rely on this indicator. It is an inexact predictor, as some does develop a bag even when they haven't been bred. In virgin does, this is called a precocious udder. Other does don't let their milk down until after they deliver, although it is not common. I have not seen this very often.

As the kid or kids move into position to be born, the doe's flanks hollow out, showing pronounced hip bones. The tail ligaments loosen so that the pelvis can widen as the kids pass into the birth canal.

THE DELIVERY KIT

The supplies you assemble to attend the birth may be as simple or elaborate as your comfort level dictates. Most of the time, a doe can deliver without help or interference. Other circumstances warrant special tools, medications, and supplements. The delivery kit list includes items that I have found useful over the years. Do not feel you need to have everything on this list. At times, my kit has been very small and didn't need to be larger. In some cases,

however, it is better to be prepared than to wish you had an item when problems arise.

THE KIDS ARE COMING

Your kidding kit is ready, and you have assembled the phone numbers of your goat mentor and your veterinarian. As the signs of kidding become more pronounced, the doe becomes more preoccupied (or starts kissing you and throwing chickens)! Probably the first sign you will see of the impending birth is a string of mucus hanging from her vulva.

Natural labor can last as long as twelve hours in first-time mothers or does with multiple kids. Normally, the serious labor immediately preceding delivery should last about an hour—no more than two. After that, carefully check to see if the doe needs help. Gloves and lubricating gel allow you to stick a few fingers into the vagina to feel for the cervical opening, the bubble, or parts of the kid.

Once the doe is in serious labor, the kids come fairly rapidly. Singles are usually born faster than twins. Triplets or more can be a problem because all those long legs get tangled up. In most cases, patience rewards you, and the kids reposition naturally as the doe paws the bedding, stretches, and gets up and down on her feet.

Delivery Kit

Basics
7% iodine
Scissors
Frozen colostrum or fresh colostrum from another doe
J-lube or other lubricant
Long, disposable surgical gloves
Nipples and bottles
Warm water for doe
Novasan or other disinfectant
10 and 20 cc syringes
20 or 21 gauge syringes

For weak or chilled kids
Kid coats
Heating pad
Blow dryer
Feeding tube and syringe

Nice to have
Baby monitor
Calendar or notebook for records
Dental floss or navel clamps

Goat serum concentrate
Kid puller or leg snare
Kid tags or ID bands
Laundry tub or kid basket
Nasal bulb
Nutridrench or energy supplement
Probiotic or yogurt

Over-the-counter medications
CMPK or calcium gluconate
Di-Methox
50% gluconate
Preparation H (relieves vulvar swelling)
Uterine bolus
C&D antitoxin

Prescription medications
Dexamethasone (for labor induction or to help premies' lung development)
Lutalyse or estrumate (for labor induction)
Oxytocin (to stimulate contractions)
Dopram (under-tongue lung medicine)

At some point, you will see a bubble appear. This amniotic sac, called the "water bag," appears and recedes as birth fluids lubricate the passage and the birth canal widens. Toward the end, the bubble should burst and release fluids. With luck, the kid's head has now crowned past the tail bones and its nose is clear so that it won't suffocate or aspirate fluid. Once the umbilical cord separates from the placenta, the kid no longer receives oxygen from the doe. The best position for the first kid to present is therefore with its nose tucked over the top of its toes.

Another normal presentation is both hind feet first, followed by the hips and tail. A second kid is often delivered this way due to the twins being nested one on top of the other, head to tail. A second common position for twins is for both to present front feet first. Sometimes kids in this position become tangled and need assistance.

Do not be alarmed if the doe cries out in distress. Most does make some sounds during labor, although some are silent. The doe may also "talk" to you in soft bleats that you will later come to recognize as goat baby talk.

After the birth, the doe delivers the placenta, also called the afterbirth. Left to her own devices, she usually eats this nutrient-rich membrane. If the delivery results in stillborn kids, however, the placenta can be invaluable for diagnosing the cause. Use gloves to collect the afterbirth along with any fetal material. Contact your vet or the state veterinary lab for help diagnosing the cause of the stillborn kid.

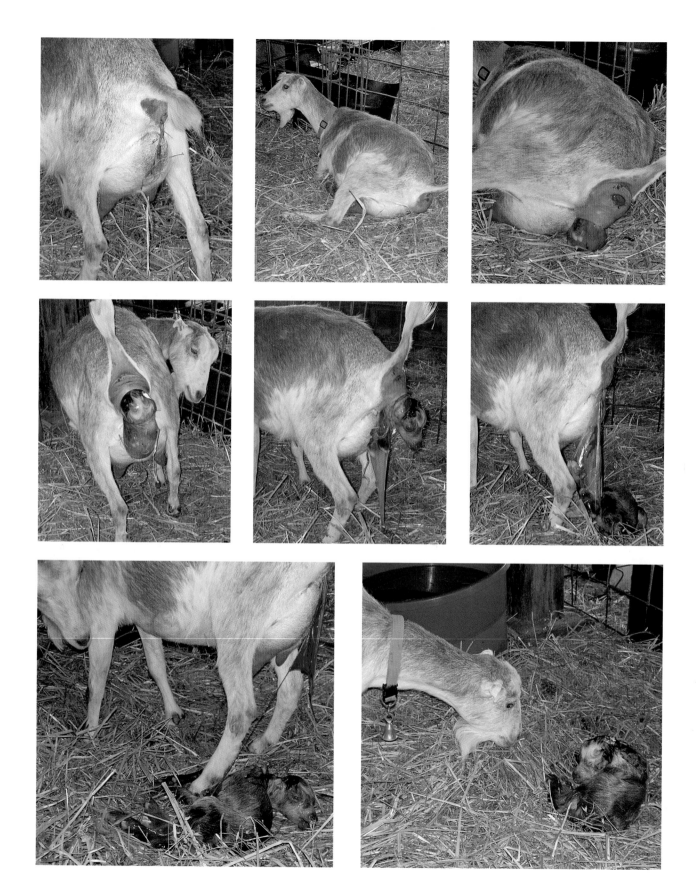

If you choose to raise your goats on a strict CAE prevention program, the birth is your time to step in. Since CAE is transmitted in bodily fluids, clean and dry the kid off with towels or newspaper as thoroughly as possible. This cleaning also mimics the mother's licking and cleaning of the kid. The stimulation arouses the kid, helping it to breathe normally. If you think it has swallowed fluids or if you find the kid still in the birthing sac, quickly clear the nostrils of membranes and mucus. It may be necessary to hang the kid upsidedown by its hind legs and gently swing it to expel swallowed birthing fluids from its lungs. Once the kid is breathing well, milk out the colostrum from the doe and save it for the kid or freeze it as an emergency supply. Before giving the colostrum to the kid, be sure to follow the directions for heat-treating found later in this chapter.

If you choose to allow the dam to raise her kid, let her lick the newborn clean and start to nurse. Constant nuzzling, sniffing, and contact are critical to forming a bond. Caprine bonding is mainly based on scent. From the moment of bonding throughout life, the doe and kid will recognize each other.

Sometimes, though, this bonding is disrupted. Kids that must be separated from the mother at birth for medical reasons but are returned within two hours will usually be accepted by the dam. After three hours of separation, the bond can be difficult to forge. Sometimes rejected kids are accepted by a doe if a strongsmelling substance such as menthol is put on the kid and the doe's nose. Resistant mothers may form a bond during forced exposure in a separate pen; however, this can take up to ten days, if it works at all.

If rejection cannot be overcome, you may find help elsewhere in your herd. Some young, subordinate does allow other kids to nurse them. Other does have very strong maternal instincts. These goats willingly raise other goat's kids that are given to them or even try to steal other does' kids.

At some point in the birthing process, you will see a bubble appear. This amniotic sac will appear and disappear as birth fluids lubricate the passage and the birth canal widens. Toward the end, the bubble should burst and release fluids. With luck, the kid's head has now crowned past the tail bones and its nose is clear so it doesn't suffocate or aspirate fluid. Once the umbilical separates from the placenta, the kid has no more oxygen from mom and needs to be out in the air as soon as possible. *Carol Amundson*

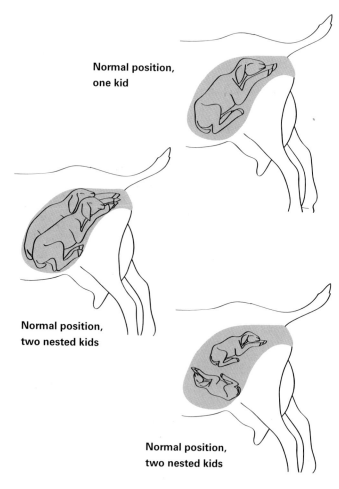

Normal position, one kid

Normal position, two nested kids

Normal position, two nested kids

ABNORMAL BIRTHING SITUATIONS

Sometimes things just don't go as planned. Intervention can be required in the case of a very large kid being birthed by a small doe, tangled kids, or a breach delivery, which occurs when the kid's backbone or tail blocks the birth canal. This type of event is when you need your vet or mentor, if possible. Delivering stuck kids is not for the faint of heart. On the other hand, with good references and courage, anyone can come to the aid of a doe.

Before placing your hand inside the animal, wash your hands with Nolvasan or another disinfectant to prevent infection. Your fingernails should also be as short as possible to avoid injuring the uterine wall. Wear gloves to help protect you from disease and to keep the inside of the doe as clean as possible. Use a good lubricant. J-Lube, a powder that becomes extremely slippery after mixing with water, is great. In very difficult cases of stuck kids, use a kid-feeding syringe to put lubricant directly into the birth canal.

People with larger hands may need to use a kid-pulling loop or a leg snare. I have no luck with these implements and prefer to feel the kid and uterus by hand. Working inside the animal is crowded. It is hard to identify the kid part or parts you are touching. You may be able to picture things better by closing your eyes as you feel inside the doe.

In spite of any fear you have that you will hurt the doe, remember that the kid is at least as large as your hand, even in miniature goats. Work slowly and cautiously. You may find that your hand is squeezed as the doe continues to have contractions. Do not be surprised if you have a bruised hand later on.

Some situations are simply beyond the goat keeper's capabilities. When necessary, call the vet for professional help and perhaps a cesarean section. Birth defects such as two-heads or kids that die before delivery may require that the fetus be cut from the doe. Luckily, such defects are extremely rare.

After assisting in the birth, be sure to clean your hands thoroughly afterward, since some of the diseases goats carry are transmissible to humans. There also have been cases of people, usually veterinarians, becoming sensitive to birthing fluids and developing an allergic response.

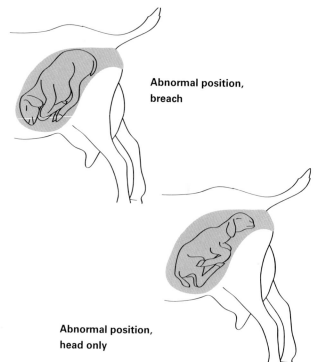

Abnormal position, breach

Abnormal position, head only

The doe licks and sniffs the kid to develop the maternal bond. *Jen Brown*

CARE OF THE DAM

Your doe has been through a difficult time. She will appreciate a warm bucket of water after delivery. Some owners add molasses to the water for energy and iron. Others give goat Nutridrench or other boosters.

If you had to assist in the birth by reaching into the doe, she'll need protection from any contaminants that may have been introduced to her womb. A uterine bolus or a broad-spectrum antibiotic can stave off infection. If she is torn or badly swollen, try Preparation H liberally slathered over the vaginal region to make her more comfortable.

The day after kidding is a good time to administer a wormer, since the stress of kidding makes any parasites the doe might have more active.

This dam is still nursing her kid at five months of age. *Carol Amundson, Cutter Farms*

CARE OF THE NEWBORN

If the kids are staying with their mom, you don't have too much to worry about. Typically, a doe cleans off her newborn quite well on her own. In colder weather or when she is having multiples, however, it is good to help her out with newspaper or old bat towels. If the weather is below freezing, use a blow dryer to prevent exposed body parts from freezing. The most vulnerable areas for freezing are the ears, tail, and feet. In my neck of the woods, there are Nubian or Swiss-type goats that we refer to as "Minnesota LaManchas" because their ears have frozen off! Angora goat kids are especially sensitive to chilling, so make sure they are warm and dry.

CARE FOR THE UMBILICAL CORD

The umbilical cord tears as the kid is expelled from the uterus. It could be any length. Use a surgical scissors to cut the cord to 2 inches.

Once you cut it—or if you choose to leave it as it tore—you should dip the cord in 7 percent iodine solution or Nolvasan disinfectant. This procedure helps prevent navel ill infection and dries the remaining tissue. While the cord dries out, the iodine also prevents bacteria from growing in the open wound.

Rarely, the cord is very thick and seems as though it won't stop bleeding. In this case, tie it off with dental floss or thread. You can also use the umbilical clamps

Kids that will be hand raised on a CAE prevention program need to be brought into another area of the barn or even the house. Laundry baskets or tubs make good nurseries for the first day or so.
Carol Amundson

After a few days, kids become active enough to jump or climb from their tub. At Poplar Hill Dairy, rambunctious kids are restrained until they can be moved to other housing. *Jen Brown*

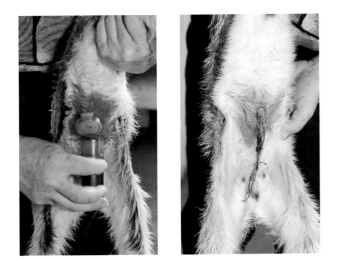

Dipping the navel with iodine or Nolvasan keeps the newborn healthy. *Jen Brown*

sold by livestock supply companies. Even when you use one of these, an iodine dip is necessary.

CHECK OUT THE KIDS

Examine the newborn kid carefully. Is it a buck or a doe? Don't look too quickly; even experienced goat keepers have been embarrassed to discover later that their newborn doe was a buck or vice versa!

Look at the jaw and palate. Very rarely a kid is born with an open palate, which is a palpable hole in the roof of its mouth. These kids can't suck properly and usually die or must be put down.

Examine the teats on both does and bucks. Ideally there should be only two. Clusters or other abnormalities may cause difficulty for the does later when milking or nursing kids. Bucks with extra or split teats should not be kept for breeding. Missing testicles or undescended testes, known as cryptorchidism, are also criteria for culling a buck. Other birth defects are always possible, so look the kids over completely.

IDENTIFICATION

If many kids are born on your farm, be sure to mark them all at birth so you know which dam had them. Marking is even important for dam-raised kids, since does have been known to steal kids from one another. In pasture or group housing, kids grow quickly. Identifying now saves confusion later.

Initial ID of kids can be as simple as a colored mark on their head or body with a matching color on the mom if they are range animals. Temporary paper or Velcro ID bands may be purchased. Breeders put the birth date, dam, sire, and other information on the band before attaching it around the kid's neck. Some goat keepers just use collars and tags right away. Tattoos stay clearer and last longer if you apply them when the kid is a few months old, so don't tattoo newborns.

INJECTIONS AND IMMUNE SYSTEM SUPPORT

Kids from unvaccinated mothers will require the C&D antitoxin to prevent intestinal problems associated with enterotoxemia. Kids should also receive a tetanus antitoxin. The antitoxins provide protection for about three weeks. In selenium-deficient areas, kids also need a Bo-Se injection.

Don't give the C&D and tetanus vaccines at this time, however, as the kid's immune system isn't developed enough to make use of the shot. Kids should be vaccinated for C&D and tetanus at one month and again at two months.

DAM-REARING VERSUS BOTTLE-RAISING

Kids can be raised by the dam or bottle-fed. Each method has breeders strongly on one side of the fence or the other. When I started raising dairy goats, I would never have considered dam-rearing except in the case of market kids. Now, after adding meat and pet goats—and finding schedule and health problems affecting the time I'm able to spend with the goats—my kid raising is strongly geared toward dam-rearing.

Dam-rearing is the least labor-intensive method of kid rearing. Most often used by the breeders of meat, fiber, or pet goats, this method requires very little attention from the owner.

Opponents of dam-raising have strong views on the matter. Dam-raised kids are said to be less people-oriented and less friendly than their bottle-fed counterparts. More importantly, from the standpoint of animal health, dams may transfer one or more of several diseases from their milk to their kids. Caseous lymphadenitis (CL),

Dam feeding *(top)* and bottle-raising *(above)* both have their merits. *Jen Brown*

CAE, and Johnes are three diseases that can pass in milk and remain latent, yet possibly infective, in the offspring. The kids run the risk of catching parasites, bacteria, and coccidia while running with the herd or even alone with their mom. Dairies may require the doe's milk for the bulk tank and prefer to find alternate sources of milk for bottle-raised kids.

On the other side, fans of dam-raising speak highly of the process, citing the fact that the kids are raised as "nature intended." Advantages include faster growth rates, fewer health problems, earlier consumption of hay and grain, and easier integration into the herd. When separated along with their dam into small pens for the first few weeks, the kids become socialized to humans as well, so they still make fine pets.

DAM-RAISING

Most goats are excellent mothers, and instinct is a wonderful thing. The best way for the dam and kid to bond is if they are allowed to be by themselves in a separate pen or a corner of the barn. It is initially important to check kids frequently to ensure that they are getting enough to eat. Once a kid stands, helping it find the nipple and even squeezing out the plug at the end of the teat gives it a good start.

If the doe is slow to let down her milk or if a kid is weak or slow to eat, be aware that it is critical that newborn goats get colostrum as soon as possible after birth, ideally within the first hour. Newborns have less than twenty-four hours to receive colostrum or their systems cannot absorb the antibodies.

If needed, you can help things along by giving fresh or frozen colostrum. Supplemental bottles do not affect the bonding between the doe and her kids. A general rule of thumb is to feed 10 percent of the kid's body weight in colostrum within eighteen hours. An 8-pound kid thus should get at least 13 ounces of colostrum.

Does with triplets may need help providing sufficient milk for their litter. When penning a group of does with kids together, do not put the smaller, multiple-birth kids and their moms with larger, single kids. Larger kids have been known to steal milk from weaker ones.

The dam bonds with the kid through scent. *Barb O'Meehan*

BOTTLE-REARING

In addition to their immediate need for colostrum, kids should get 15 percent of their body weight in milk each day. This is ideally divided into three to four feedings timed equally apart. Some breeders find that twice-daily feedings are enough. The kid's tummy is a good indicator. When it is obviously full and slightly distended, he has had enough—no matter how much he thinks he should keep drinking. Kids should be weaned at eight to twelve weeks, once they are eating hay and chewing a cud. Make fine-stemmed grass hay and fresh water available all the time from about one week of age.

Leaving kids in pasture lets them play in the sun.
Barb O'Meehan

Multiple kids can eat from a Lambar, which is a bucket with nipples and tubes going into the milk *(top)*. As with bottle feeding, kids need to be trained to use the nipple *(above)*.
Carol Amundson (top), Jen Brown (above)

CHAPTER 9

• • • • • • • • • • • • • • • •

KEEPING HEALTHY GOATS

I don't know how old I am because the goat ate the Bible that had my birth certificate in it. The goat lived to be twenty-seven.

—Satchel Paige

This chapter discusses the pros and cons of management techniques, plus techniques for some of the most common practices. Basic management decisions you'll need to make for your herd include how to handle hoof trimming, basic grooming, horns, neutering, parasite prevention, and vaccinations. Everyone has different ideas about what practices are necessary based on their management style, budget, and personal experience. My own goats are disbudded, vaccinated, wormed, and trimmed on as regular a schedule as time permits. Unwanted males are neutered as well.

BIOSECURITY

The word biosecurity calls to mind images of large, antiseptic factory farms. However, it applies to all operations, large or small, and can be the key to keeping healthy livestock. Any measure that decreases the

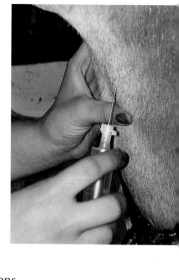

chance of disease-causing organisms entering the goats' environment increases your level of biosecurity. Each goat keeper needs to determine how far to take the following biosecurity measures:

- Clean and disinfect pens frequently.
- Limit or restrict bringing outside animals into the herd.
- Isolate new animals, sick goats, and those with chronic contagious conditions.
- Limit or eliminate any travel for sick goats, including shows.
- Change your clothes or put on coveralls when moving from regular pens to isolation pens.
- Wash your hands frequently between animals and before going from pen to pen.
- Change and disinfect your clothes and foot coverings when you come home from shows or other farms.
- Restrict visitor access to livestock housing and limit their handling of your goats.
- Request that visitors do not wear the same clothes from their barn to yours. Have a change of outer clothes or boots for them.
- Use gloves for invasive procedures or those involving body fluids.
- Disinfect or change equipment between goats when performing procedures that involve needles, tattoo digits, hoof trimmers, or other tools that come into contact with body fluids.
- Properly dispose of potentially biohazardous materials.

BLOOD TESTING

Goats may be tested for a variety of diseases and conditions. Buyers or owners may request testing for specific conditions like CAE to facilitate purchase or culling decisions. Your veterinarian can perform the blood collection. However, many breeders who use regular

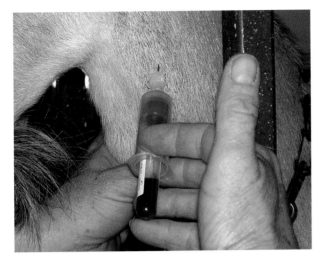

When performing caprine venipuncture, you need to secure the vein *(opposite)* and then fill the tube *(above)*.
Carol Amundson

testing as part of their herd management find it less expensive and less troublesome to collect their own blood samples.

HOOF TRIMMING

Hooves are the foundation of the goat. The nimble movements and breathtaking climbs that goats love to perform are impossible with overgrown, cracked, or sore feet. Most goat owners dislike the chore of hoof trimming and avoid it whenever possible. But it is one of the nicest things you can do for your goats—even if they don't believe it.

Standard schedules recommend trimming hooves every three months. In reality, the time between trims should be based on each individual goat. Goats that spend time on hard ground and rocks wear their hooves down naturally. When goats are kept on soft bedding, their hooves grow rapidly. Goats that spend time in wet areas collect "crud" between their toes and in the overgrown part of the hoof, which can lead to foot rot.

The main tool needed for trimming hooves is a pair of good trimming shears or a knife. A hoof pick can be useful for cleaning out crud. Some people "sand" the hooves as a finishing touch using a hoof file or hoof plane. Many goat keepers recommend shears because a knife is too difficult to use on tough hooves. After trying a number of products, I discovered the "magic" orange-handled shears. These Teflon-coated shears, available from livestock supply catalogs, stay sharp and do not fall apart easily. Ergonomically designed, they allow more trimming without causing hand cramping or blisters.

Caprine Venipuncture

Materials

Livestock clipper
Alcohol wipes
Vacutainer and holder
18-21 gauge needle
Specimen tube (usually red top for serum testing)
Pens and labels for tube

1. Shave the neck over the jugular vein.
2. Cleanse the area with alcohol.
3. Secure the vein with your thumb so that it stands out.
4. Insert needle at a 45-degree angle. Use a quick, smooth puncture to reduce pain and trauma to the goat.
5. Hold the needle steady while filling the tube.
6. When the tube is full, either fill a second tube or remove the needle from the goat.
7. Identify sample with at least two forms of ID. Useful information for labeling a tube could include the goat's name, the tattoo or ear tag, registration ID, date of birth, or the name of the owner or herd.

Horn Prevention or Removal

TECHNIQUE	COMMENTS
Disbudding	• Recommended method for caprine horn prevention. • Perform when kid is 3 to 7 days old or when horn buds are found. • Take care not to overheat the brain. • Give tetanus antitoxin when immunity is questionable.
Banding	• Removes small to medium horns. • Flies can be a problem during fly season. • Pain may occur when nerve is severed. • Scurs may replace horn when bud isn't completely destroyed. • Sometimes bloody. • Requires a disbudding iron or heated rod for cauterizing bleeds. • Works on some goats, unsuccessful on others.
Veterinary removal	• Costly. • Difficult to find vets experienced in horn removal. • Vet can provide pain blocks or anesthesia during procedure.
Sawing	• Bloody. • Cauterize bleeds using a disbudding iron or heated rod. • Scurs may replace horn when bud isn't completely destroyed. • Traumatic and painful to the goat. • Saw, especially wire blades, can slip and cause injury to you.
Caustic paste	• Failure is likely. • Hazardous to goats' eyes. • Kids rub or lick paste off pen mates, causing burns. • More commonly used on cattle; not recommended for goats.
Gouging	• Bloody. • Device is used to scoop out the horn buds. • More commonly used on cattle; not recommended for goats.

for Bridget, one of our horned Boers that gets stuck on a regular basis.

There is a very real possibility for horned goats to injure people. Much as we'd like to believe that all goats are raised properly, some learn to butt people. This behavior can be painful and annoying from a goat with no horns. From a horned goat, butting can injure or kill someone. A swinging goat horn is just the right height to accidentally hit a child in the eye.

Use caution when mixing horned and dehorned goats in the same herd. Make sure the horned animals aren't too aggressive. Goats naturally work out a hierarchy, or pecking order. They can be brutal about using their horns on other goats to establish their position in the herd.

DISBUDDING

The best way to remove horns is to catch them before they develop. Large buds and small horns that have already formed take longer to disbud, which is more traumatic for both the goat and the handler. When performed too late or too lightly, disbudding fails to completely destroy the horn bud. Incomplete horn removal results in scurs as the animal ages. Scurs are deformed horns.

Disbudding is one of the hardest tasks for the new goat owner to learn, but it is crucial in most herds. I suggest that you find a goat owner who is willing to teach you the process the first few times. I work with new goat owners all the time and consider it a duty to pass along the skill I learned from the person who sold me my first animal.

The best age to disbud is between three and seven days old. The kid must be strong on its feet and have a palpable horn bud. The timing is subjective; some kids (especially doelings) are slow to develop their horn buds, while some bucklings are born with the buds already starting to grow.

Disbudding is done with a hot iron specially designed for dehorning goat kids (I use the Rhinehart X50). First, use a hair clipper to remove hair around the bud. Trimming prevents the strong smell of burning hair and helps you see very small horn buds.

This step is especially helpful to new goat owners. In larger operations, hair clipping is typically considered unnecessary and time-consuming. An inexpensive wire brush from the hardware store cleans the excess hair and other burned debris from the tip of the iron to make it more efficient. While I am guilty of forgoing gloves, it is best to use them to prevent burns to yourself. A disbudding iron is very hot and burns at a simple touch. (I have the scars to prove it!)

A disbudding box is a simple structure used to restrain the kid during bud removal. Oftentimes, it's just as easy to disbud without the box, and I use mine only when the kid is too unruly.

If the doe was vaccinated for tetanus before kidding, the kid will have acquired immunity from its mother. If the vaccination status of the dam is unknown, give the kid a dose of tetanus antitoxin at the time of disbudding.

Some breeders use topical antiseptic sprays or numbing sprays during disbudding to stave off infection and provide comfort. Solarcaine or other sunburn sprays seem to provide relief, if only for the goat owner. Choose a product with the highest level of lidocaine and apply immediately after disbudding to cool down the head. It isn't helpful to apply spray before the procedure, as the numbing effects are minimal and the spray is burned off in the process of disbudding.

Late or improper disbudding can result in deformed horn growth called scurs. *Jen Brown*

The best way to eliminate horns in your herd is to disbud kids within the first week of life. *Shutterstock*

An adult goat can be dehorned by banding or sawing, but this is traumatic to the goat and increases the risk of infection.
Jen Brown (above, left), Carol Amundson (above, right)

The initial chance of infection from disbudding is slight because the hot iron seals the bud. Greater chances of infection arise later, when the head starts to heal and the scabs get knocked off, which can leave a slightly bloody wound. A furazolidone spray (Furox) or other wound spray can also help, either at the time of disbudding or when there is bleeding from the scab.

While disbudding, it is okay to go back and forth from one bud to the other to allow the spot to cool a bit. The temperature of individual disbudding irons varies greatly. Know your iron. Remember that you only want to burn enough to remove the bud; too much burning will overheat the kid's brain or break through the skull. Be aware of the kid at all times when disbudding rather than relying strictly on a time count for the process.

Older irons wear out, so watch the condition of your equipment. One herd manager who had disbudded hundreds of kids without incident suddenly had a number die after disbudding. After examining the kids and the iron, he discovered that the iron had worn thin and the hot, sharp metal was piercing the skull, causing severe brain injury. This is an extremely rare occurrence but bears mentioning, especially if you purchase used equipment.

LIVESTOCK GUARDIANS

Goats can be the victims of accident, theft, or predation. A livestock guardian (LG) is a good investment, particularly when the goats reside a distance from the main farm. Most commonly, the guardian is a dog, but

Clean copper-colored rings indicate a successful disbudding.
Jen Brown

llamas and donkeys are also used. While not a cure-all, the presence of a guardian can deter predators. The true guardian that runs with the herd is a special animal. Personality and training both factor into its effectiveness. Look for breeders who raise guardian animals, and remember that even good livestock guardians need to be trained to respect the stock.

Livestock guardian dogs (LGDs) have been used for centuries in Europe. The common breeds originated there, often carrying the names of their region, including Great Pyrenees, Akbasch, Maremma, Anatolian Shepherd, and Komondor. Some, like the Great Pyrenees, prefer a large territory.

Guardians need to be capable of independent thinking. Overly shy animals should be avoided. Livestock guardians are different from herding dogs in that they do

Disbudding

REQUIRED
- Disbudding iron
- Wire brush

OPTIONAL
- Small hair clipper
- Disbudding box, homemade or purchased
- Gloves
- Tetanus antitoxin
- Solarcaine or other burn relief spray

1. Preheat your disbudding iron. Be sure to avoid placing it in a spot where it may start a fire or where the cord will be tripped on by goats or people. This tool gets VERY hot!
2. While waiting for the iron to heat, give the dose of tetanus antitoxin (unless the kid already has immunity from its dam or an earlier injection).
3. Restrain the kid in a box or on your lap.
4. Clip the hair from over the buds.
5. Check to see that your iron is hot enough. The tip will glow red if you are in a darkened room. The tip should leave an immediate round burn scar on a piece of wood.
6. Set the tip of the hot iron directly over the center of the horn bud.
7. Apply gentle, even pressure while twisting the iron back and forth in a circular motion to help provide an even burn. Some people count to ten slowly or repeat the phrase "This will save your life" ten times.
8. Raise the tip of the iron to see if the burn is complete.
9. Remove the burnt skin, hair, and ends of the loose horn (especially important for bigger buds on older kids).
10. Reburn as necessary until there is a clean, copper-colored ring around the bud.
11. Repeat on the other bud.
12. Spray head with burn relief as desired.
13. Give the kid a little extra petting or a bottle and release.

Jen Brown

not have the instinct to chase. While younger dogs may do some chasing, the older dog is more sedentary, preferring to hang out with the herd. You do not want a dog that is aggressive toward people or prone to wandering.

Training depends on your needs and the nature of your farm. Animals trained to range are frequently restricted from human contact at eight to twelve weeks of age to facilitate bonding with the goats. Dogs on smaller farms should be more socialized to people. Many of these large dogs are very calm and seem almost lazy as they lie about in the pasture during the day. At night, they go into action, patrolling boundaries and barking warnings. This barking may be a problem for small goat farms that are in more residential areas with close neighbors.

Expect to conduct extensive training of your dog, including teaching it not to chase or roughhouse with the goats. These dogs mature slowly and do not leave puppyhood before the age of two or three years. Immature dogs should be kept in smaller areas and monitored.

Unfortunately, the large-breed canines are not long-lived animals, often having an expected lifespan of eight to ten years. Consider adopting a younger dog as the old guardian ages so that the youngster may be trained by a canine mentor.

Livestock guardian dogs require strict training to be effective. They must respect the goat herd.
Carol Amundson

Llamas and donkeys use many of the same feeds and medications as goats, which make them easy to manage as livestock guardians. Both of these animals have an instinctive dislike of canines. They will drive off intruders by calling out, kicking, and chasing the stranger. Intact males are not usually used due to their aggression toward goats, including a tendency to mount or attempt breeding. Most often, a single llama or donkey is used in the herd since, like a dog, the guardian needs to bond with the herd, not with people or others of their species.

NEUTERING UNWANTED MALES

Neutering isn't a pleasant task. However, neutering eliminates the characteristics that make bucks objectionable, such as odor and breeding behaviors. Wethers make excellent companions or working goats. In meat herds, wethers reduce rut behaviors. Fiber goat wethers have cleaner fiber. Neutering can also extend the goat's lifespan, since wethers do not go through the stress of rut every fall, which tends to make bucks shorter-lived on average compared to other goats.

Bucks are usually born with their testicles descended. While neutering is possible even at a few days of age, some breeders choose not to castrate until the animal is eight to twelve weeks old. This allows the urethra time to develop and may help prevent urinary calculi problems later; although other reports indicate that the age at neutering is a less important factor than the animal's diet.

Ethical and market considerations should be taken into consideration when deciding on your farm's neutering philosophy. In meat animals, some buyers require intact animals, while others prefer wethers. Some sellers do not allow any non-breeding male to leave the farm without neutering. This keeps unsuitable genetics from entering the gene pool as well as preventing new owners from unwittingly purchasing a pet goat that later turns into a stinky, unpleasant menace.

Methods of neutering include cutting the scrotal sac and removing the testicles, crushing or "crimping" the spermatic cord with an instrument known as an emasculator (Burdizzo), and banding the testicles with an elastrator band.

On our farm, we typically castrate by cutting the male kids at a young age when their testicles are still small. Banded kids seem to be in distress longer than those that are cut or crimped. Some European countries consider banding inhumane and have made it illegal. Because the scrotal sac actually rots off, this technique has the highest risk of infection.

Some breeders give pain medication prior to performing castration or other traumatic procedures. Oral pain meds are poorly absorbed by goats, but a regular aspirin given at the dose of one tablet per 10 pounds or oral Banamine at 1 milligram per pound may be given a few hours before the procedure. An injection of Banamine can be procured from a vet and given a half hour before cutting or crimping. Fias Co Farm in Tennessee sells an herbal painkiller known as Owe-ese on its website, and some other breeders use willow branches for the natural salicylates that act similarly to aspirin.

Neutering

METHOD	PROS	CONS	SUPPLIES	SAFE AGE
Banding	• Inexpensive. • Bloodless.	• Least humane method. • Technique error may leave a testicle and fail to sterilize. • Risk of tetanus.	• Elastrator. • Bands or rings. • Tetanus antitoxin if not vaccinated.	• After the testicles descend. • More traumatic the larger the testicles.
Cutting	• Inexpensive. • Most reliable method.	• Open wound. • Risk of tetanus. • Bloody, may be disturbing. • Risk of excess bleeding in older kids.	• Surgical scissors or scalpel. • Disinfectant. • Tetanus antitoxin if not vaccinated. • Pain medication if desired.	• After the testicles descend. • More traumatic the larger the testicles. • Do not cut kids with a scrotal hernia. • Bucks over 6-8 weeks should be done by a vet with anesthesia.
Emasculation	• Quick recovery. • No cutting or blood involved. • No chance of infection. • Relatively humane.	• High initial equipment cost. • Technique error may result in incomplete castration.	• Burdizzo or Nipper. • Pain medication if desired.	• 4 weeks or older. • May be used on full-sized bucks.
Vasectomy	• Useful for teaser buck. • Very reliable.	• Expensive. • Leaves breeding traits intact.	• N/A	• Any age.

Most large herd operations castrate their own male kids, but veterinary help may be preferable for specific cases or smaller herds. The vet has sedation available. Full-size bucks may be neutered by crushing the spermatic cord with a Burdizzo. We have had this done by our vet rather than attempting it ourselves. Veterinarians may also perform a vasectomy if you want a teaser buck for the herd. This method does not eliminate breeding behaviors, so it's unsuitable for pet goats.

Always begin any home neutering operation by washing your hands and the instruments thoroughly with soap and water. Also wash the buckling's scrotum,

When talking parasite control, goat owners most often mean internal parasites. Coccidia are microscopic organisms. Worms such as the roundworm and tapeworm are larger. These creatures live in the caprine digestive tract, taking nutrients from the intestinal contents or sucking blood from the animal. More difficult organisms to treat are those that live outside the intestines and do not show up during fecal exams. Liver flukes and lungworm are examples, as is the deer worm, which attacks brain and nervous tissue.

As with other management tasks, approaches to parasite control range from nonintervention to regular testing and treatment, as well as automatic, timed treatment without testing. Unfortunately, some wormers are becoming ineffective as parasites build up resistance. Many wormers now reduce parasite loads by fewer than 5 percent after treatment.

Clinical signs of parasitism include rough hair coat, diarrhea or soft feces, swelling under the jaw, and pale membranes in the inner eyelids. But the best first step in establishing effective parasite control is to test your goats. Identifying a baseline of pest levels will help you determine which wormer will be most effective. Owners can take individual goat fecal samples or a mixed herd sample to their vet for a worm count.

Meat-goat farmers in the United States have started using the FAMACHA© parasite control system developed in South Africa. This method measures anemia, useful in detecting the roundworm *Haemonchus contortus*. This method must be used in hot weather, when parasites are active. The bottom eyelid of the goat is pulled down and the color of the membrane is compared to a color-coded chart, which follows a scale from 1 (healthy color) to 5 (anemic and pale). Herd owners using FAMACHA© should still check fecal specimens every few years to confirm the effectiveness of their parasite control program.

Coccidia are more of an issue in kids, younger animals, or those with poor immune systems. Older goats have resistance to these microscopic protozoa. The majority of coccidia are specific to goats. Two exceptions are cryptosporidium and toxoplasmosis, which can infect other animals as well as humans. Kids in larger herds are often treated automatically. Over time, the goats develop resistance to these pests. Coccidostats for kid-rearing are some of the most frequent wormers given on an automatic basis.

VACCINATIONS

Vaccine management differs between conventional and holistic approaches and can be a matter of debate. Vaccines are not harmless, so most goat keepers vaccinate only if the threat is real.

There are two types of immunity. Passive immunity, which is transmitted from the dam, creates healthy kids. Maternal antibodies pass through the placenta to kids in

Guidelines for Parasite Prevention

- Avoid contamination of water buckets by animal or bird droppings.
- Place feed off the ground to keep feet and manure out of the feed sources.
- Maintain low stocking rates: no more than six to eight small ruminants per acre. Alternate grazing of cattle or horses can help.
- Rotate grazing areas and pens.
- Sanitize kid-rearing pens and equipment between batches of kids.
- Select medications and dosages based on herd observation and worm load testing.
- Treat with drugs based on need or FAMACHA© scores, dosing only goats with high parasite loads.
- Withhold feed for twelve hours prior to administering oral wormers.

Vaccination Considerations

- Preferred management style:
  - Conventional
  - Sustainable
  - Organic
- Risk tolerance for disease.
- Known risk of exposure versus risk of adverse reactions.
- Effectiveness of the vaccine.
- Safety of the vaccine.
- Benefit of the vaccination versus the expense.
- Necessity of a yearly booster.

utero. Colostrum provides antibodies that protect kids for the first two months of life.

ARE VACCINATIONS NECESSARY —AND WHICH ONES?

Most goat owners follow some type of vaccination program. Some vaccines are given only when the disease is present in the herd and after management changes have failed. Keys to successful vaccination include proper storage and handling of the vaccine and care to follow recommended inoculation times and doses.

Advocates feel that vaccinations are a form of cheap insurance against disease. The most commonly recommended caprine vaccine is CD/T, which provides three-way immunity against clostridium types C and D, which cause enterotoxemia, and tetanus.

Compared to human vaccine schedules, our pets and livestock receive boosters considerably more often than we do. Vaccine detractors feel that some (or all) vaccines are unnecessary and that yearly booster shots are too frequent. Some experts believe that annual revaccination is

unnecessary because the initial series of shots provides long-term immunity.

RABIES VACCINATION

The risk of rabies in domestic goats is slight. No rabies vaccine has been licensed for goats, although the large-animal rabies vaccine is approved in sheep. It is a good idea to vaccinate all dogs and cats on the farm for rabies. Consult your veterinarian to decide if it is necessary for your goats.

OTHER VACCINATIONS

There are many other vaccines on the market for diseases such as bluetongue, chlamydiosis, *E. coli*, foot rot, Pasteurella, and sore mouth. Some of these diseases are regionally troublesome, so local practitioners know best whether there is a need for vaccination. Check with a veterinarian to be sure the disease and organism is found in goats before challenging the caprine system with an unnecessary shot.

Criteria for Culling

- Bad bite, broken or missing teeth, problems eating.
- Bad teats (too big or too small).
- Bad testicles (split, too small, infected).
- Bad udders (lopsided, poorly attached), especially in dairy goats.
- Does that have not settled for two seasons.
- Evidence of abscesses or other disease.
- Poor body condition due to illness or old age.
- Poor-quality fiber on fiber goats.
- Structural defects (bad feet, legs, back).

CHAPTER 10

DEALING WITH
SICK
GOATS

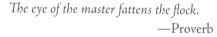

A sick goat is often listless and uninterested in her surroundings. You can tell just by looking at her that this goat is ill. *Carol Amundson*

The eye of the master fattens the flock.
—Proverb

Nothing strikes terror into the heart of an animal lover more than illness or injury to one of his or her charges. This topic alone fills books. A number of excellent resources are devoted to the medical care of goats.

Goats are susceptible to the same types of health problems as any living creature. Being in our care, they deserve proper treatment during illness. Basics include a dry, clean environment, food and water, and safety from harassment—perhaps a separate pen—when they do not feel well. Because a herd always has a pecking order, sick animals may get harassed by lower-ranking herd mates.

Treatments range from holistic to traditional. I have seen goats left alone and seemingly dying that have recovered to live many more years. I have also attentively nursed goats for weeks using the best care available, yet they still died.

In the strictest form, organic practitioners do not use antibiotics or other chemical treatments. Other people put as much medication into the goat as they can afford. I favor a middle ground.

OBSERVATION AND EXAMINATION

Spending time with your goats each day is the best way to keep ahead of health problems. Goats instinctively try to hide an illness as long as they can. Knowing your goat's normal behavior will help you detect problems early on. If something seems off with your goat, it very well could be.

TEMPERATURE

The normal temperature of a goat ranges from 101.5 to 104 degrees Fahrenheit. Caprine temperature varies with the weather as well as by type of goat. Heavy hair coats make a goat warmer in hot weather. The best way to know what is normal in your herd is to check the temperature of several goats so that you can tell how the average animal is running.

A high temperature indicates infection, often bacterial, which means that your goat may need antibiotics. A fever is part of the body's natural defense as it tries to "burn out" the offending agent. Certain medications can help bring the goat's temperature down. If the high temperature stems from an untreated infection, however, the cooler temperature could allow the infection to worsen.

A digital thermometer is the easiest to read and not very expensive. Be sure to get a rectal thermometer. Use a

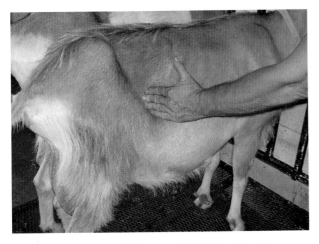

Measuring rumen motility is a useful skill for detecting abnormalities in your goat's digestion. *Carol Amundson*

light coating of lube or oil. Holding the goat's tail, insert the thermometer end into the anus. Hold the thermometer in place until it signals that the temperature has been taken. Remove, read, and record the goat's temperature.

Temperatures below normal occur in critically ill animals and chilled newborns. Return the body to its normal temperature with heating pads, heat lamps, or even a hot bath. To warm a chilled kid, wrap the kid in a plastic bag to keep it dry and submerse it to its neck in hot water.

RUMEN MOTILITY

The rumen is a major organ in the goat. Rumen disorders are the root of most critical caprine illnesses. Digestive upsets can become life-threatening without warning. Knowing how to check rumen motility, or movement, is a useful skill.

The normal rumen contracts one to four times per minute. Less frequent motility occurs when the stomach is at rest with very little recent food inside it. Faster motility occurs after a meal. To measure rumen movement:

1. Put your fingers on the left side of the goat, between the ribs and hips in the soft hollow below the loin.
2. Feel for a hard mass—that is the rumen's contents.
3. Hold your fingers in place until you feel a rolling contraction.
4. Start timing until you feel the next movement.
5. Count the contractions for 1 minute.

Caprine Vital Signs

VITAL SIGN	NORMAL RANGE	NOTES
Heart rate	70–80 beats/minute	
Temperature	101.5–104°F	• Varies with environmental temperature and activity. • Lower in the morning. • Test healthy goats in herd for a benchmark.
Respiration	12–15 breaths/minute	
Rumen motility	1–4 movements/minute	• Faster after a meal. • Slower if stomach is empty.
Rumen pH	5.5–7.0	

Goat Health Check

OBSERVATIONS	SIGNS OF HEALTH	SIGNS OF ILLNESS
Appetite Water usage	• Appetite normal. • Interested in food. • Drinking normal.	• Won't eat or drink. • Too much interest in food. • Drinking too much water.
Attitude Alertness	• Bright and alert. • Inquisitive. • Normal behavior.	• Hunched back. • Moaning or crying. • No interest in surroundings. • Staring into space. • Tail drooping. • Tremors or shaking. • Unresponsive.
Body condition	• Body condition good.	• Too fat or too thin.
Ears	• Normal ears.	• Shaking head. • Drooping ears. • Visible parasites or discharge.
Eyes	• Clear and bright. • No discharge. • Able to see.	• Cloudy or discolored. • Sunken, squinting, or shut. • Discharge, tearing. • Blindness.
Feet, hooves, legs, joints, and gait	• Stands comfortably. • Moves easily. • Puts equal weight on feet.	• Pain or swelling. • Limping or lameness. • Unwilling to stand.
Lymph nodes	• Normal.	• Swollen or lumpy.
Manure	• Normal pellets.	• Pellets too dry. • Watery stool or mucus present. • Feces bloody.
Mucous membranes (eyes and gums)	• Pink. • Moist.	• Pale. • Dry. • Red or off-color.
Respiratory	• No abnormal sounds. • Clear or no nasal discharge.	• Rasping breath, rapid breathing. • Abnormal cough. • Green or cloudy nasal discharge.

OBSERVATIONS	SIGNS OF HEALTH	SIGNS OF ILLNESS
Skin and coat	• Skin supple. • Smooth, silky coat.	• Skin dry and flaky. • Coat dull or hair falling out. • Wounds or lumps.
Teeth and mouth	• Teeth good. • Breath normal. • Mouth and tongue normal.	• Teeth missing or broken. • Grinding teeth. • Breath smells abnormal. • Scabs or sores. • Swollen tongue.
Udder	• Normal shape and texture. • Normal production. • Milk white and sweet.	• Abnormally swollen or hot. • Sudden drop in production. • Milk bloody, gassy, or watery. • Off-tasting milk.
Urine	• Normal color and amount.	• Blood or crystals. • Visible dribbling or discharge. • Straining.

MEDICINE

Goats are a minor species in the farming world and often placed in the generic category of small ruminant, alongside sheep. Due to economics, drugs are rarely tested or labeled specifically for goats. Some medications, most notably wormers, must be given to goats at double the dosage prescribed for other species to be effective. Depending on who you talk to, a particular drug may be deemed very effective for goats or useless. Along with recommendations from your veterinarian and advice from other goat owners, your observations will help you decide what works best for your herd.

There are several recipes and treatments you can make to benefit your goat's health from supplies in your garden or kitchen cupboard. I have collected these recipes over the years and find them helpful in a pinch.

ELECTROLYTE REPLACEMENT FLUID

When a goat refuses to eat, it may become dehydrated and its electrolyte balance could be disturbed. This can also happen when the goat has extreme diarrhea.

1 gallon warm water
2 teaspoons table salt
1 teaspoon baking soda
½ cup honey, Karo syrup, or molasses
    (never cane sugar)

Allow the animal to drink free choice or drench (as described on page 112) conscious animals with small amounts at frequent intervals until interest in food and water returns.

Goat Medicine Chest

- Alcohol or alcohol preps
- Bandage materials
- Drench gun or syringe
- CMT test kit (to test for mastitis)
- Castrating supplies (surgical scissors, bander, or Burdizzo)
- Clippers and supplies for clipping
- Collars and leads
- Disbudding iron
- Eye puffer or ointment
- Electrolyte replacement powder or fluid
- Fecal test kit and microscope
- Feeding tube for kids
- Gloves, leather
- Gloves, surgical, long and short
- Hoof trimmers
- 7% iodine solution
- Needles—21 or 22 gauge × 1"
- Measuring tape or weight tape
- Nolvasan (residuals said to last for two days) or disinfectant
- OB lube (J-Lube powder)
- Peroxide
- Scissors, bandage and surgical
- Splints
- Stanchion and head gate
- Stomach tube for adult goats
- Syringes, 3cc and 12cc
- Thermometer
- Udder infusions
- Vet wrap
- Wound ointment or spray
- Weak kid syringe and stomach tube
- Wormer

MAGIC SUPPLEMENT

Another homemade potion, Magic is a quick energy supplement. Originally, I made this for does who had just kidded as a pick-me-up or to treat early ketosis. Magic helps with iron replacement and energy. It can be given in any situation where the doe is rundown and in need of nutrients.

1 part molasses
1 part corn oil
2 parts Karo syrup
Dose: 120cc two times a day

WOUND OINTMENT

This ointment can be used in place of many of the wound sprays available from livestock suppliers. It covers the gamut from bacterial protection to ringworm treatment and is very soothing.

1 medium container of Vaseline
1 large tube of diaper rash ointment
1 tube of women's yeast infection medication
1 tube athlete's foot medication
¼ cup Nolvasan or Betadine liquid
1 tube triple-antibiotic wound ointment

Warm gently to liquefy, blend, and allow to cool.

INJECTIONS

Many medications and vaccinations are given by injection. Daunting at first, this simple procedure can be learned by any goat keeper. Shots given in the muscle are called intramuscular (IM); shots given under the skin are called subcutaneous (sub Q). An increasing number of injections these days are sub Q. The goat can absorb medication from a sub Q injection without the muscle damage and scarring of IM shots—especially important in meat goats.

Most farmers give shots behind the elbow, where goats do not have lymph nodes. This location prevents injection lumps, which are caused by reaction to the shots, from being mistaken for CL lumps, which occur

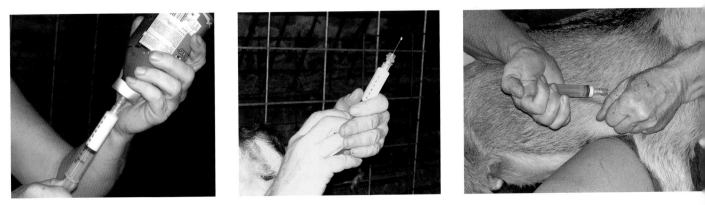

To give medication by injection, you need to fill the syringe until you have the proper dose *(above, left)*, remove air from the syringe by gently depressing the plunger until a drop of liquid appears on the end of the needle *(above, middle)*, and inject medicine slowly into a small tent in the skin *(above, right). David Weber, Cutter Farms*

Adverse Reactions to Shots

Sometimes a goat has an adverse reaction to an injection of medication or supplement. Epinephrine can save a goat's life when the reaction is severe enough to cause anaphylactic shock. Epinephrine must be given at the first sign of a reaction, so keep a bottle handy.

REACTION	CONCERN	TREATMENT
Lumps and swelling of injection site	• Pain. • Unsightly. • Possible to confuse with CLA lumps.	• Rub injection site gently after giving the shot. • Ice the swelling. • Give anti-inflammatory medicine. • Always inject behind the front leg.
Lameness		• Give anti-inflammatory medicine
Dragging leg	• Paralysis of the nerve. • Usually associated with penicillin given in the hind leg.	• Give anti-inflammatory medicine. • Avoid injecting goats in the hind leg or rump.
Rash or raised bumps seen as little spots of raised hair all over goat	• Allergic response or mild anaphylaxis.	• Give epinephrine. • Give Benadryl.
Difficulty breathing Trembling Sudden collapse	• Anaphylactic shock.	• Give epinephrine. An injection under the tongue works the fastest in the case of a severe, immediate reaction.

An oral syringe can be used by itself or attached to a stomach tube. *David Weber*

To administer a drench, the dosing syringe is inserted into the goat's mouth, usually with some objections from the goat. Make certain the goat is swallowing the liquid and it isn't entering her lungs. *David Weber, Cutter Farms*

in the lymphatic system of diseased goats. This is an important consideration for showing goats.

A sub Q injection is given by pulling up a little pinch of skin into a "tent." The needle is inserted into the side of the tent. Take care not to go through the skin on the other side. A gentle pullback on the syringe plunger will tell you if you are under the skin and can go ahead with the shot. Air in the syringe means you have gone through the skin and must reposition the needle.

An IM injection is usually given into the large muscle in the lower hip; the shoulder is a secondary site. Goats should not be given shots in the rump. To be certain you are not in a blood vessel, pull back the plunger and check

for blood in the syringe. Incorrect injection into a vein can be hazardous and may even kill the goat, so repositioning the needle is important when blood appears.

ORAL MEDICATIONS

Oral medication may be administered in food or water, as a drench, or as a bolus. A drench is liquid medication that is administered down the goat's throat. A bolus of oral medication is a large pill.

The easiest method, of course, is mixing the medication with food or water. The ailing goats have no choice but to consume it or go without food or water. The problem is that this method is imprecise. Some goats inevitably get too much medication while others do not receive enough.

Administering oral medication directly to the goat assures proper dosage, but only if you get the medicine inside the goat! Getting a liquid drench down the goat's throat—as opposed to all over you, the goat, and the floor—can be tricky. And trying to insert a pill into a goat with your hand is a great way to get bitten by its sharp back teeth. Goats are adept at resisting pills. Usually, after dislodging the offending hand, the goat will then contemptuously spit the pill at your feet!

To combat these issues, restrain the goat. Options include holding the goat firmly between your legs, having a helper hold the animal, or using a head-gate.

To administer a drench, you will need a commercial drenching gun or a simple needle-less syringe. The technique is the same with either implement. To administer a bolus, you need a balling gun. This implement is available from vet suppliers. It comes in several different sizes, the largest of which works better for cattle than goats. The balling gun gets the pill far enough down the animal's throat to force it to swallow. Care must be taken not to accidentally place the pill into the animal's lungs.

STOMACH TUBES

Drenches should only be used in animals that are alert enough to swallow. Weak or unconscious animals need a stomach tube to prevent aspiration of liquid into the lungs. Most often, a stomach tube is used with young kids, especially to give colostrum to weak kids or those that will not take a bottle within the first few hours after

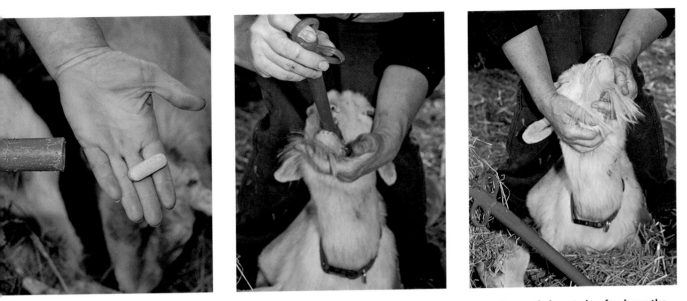

A bolus is a large pill administered with a balling gun *(above, left)*. The balling gun needs to be gently inserted as far down the goat's throat as possible, then the plunger is depressed, releasing the pill *(above, middle)*. Afterward, hold the goat's mouth closed and rub its throat so it will swallow *(above, right)*. *David Weber, Cutter Farms*

delivery. This extremely useful tool consists of a flexible catheter and a large syringe. For adult goats, livestock supply houses carry large-animal stomach tubes. A large syringe or funnel is used to deliver the fluid.

It is important to be certain you have threaded the tube down into the stomach rather than the lungs. Aspiration pneumonia or drowning can occur when the lungs are suddenly loaded with liquid.

How to Use a Stomach Tube

1. Measure the tube against the outside of the goat from nose to stomach to estimate how far it needs to be inserted. Mark the tube.
2. Restrain the goat and hold its head upright. For older goats, put a short length of polyvinyl chloride (PVC) or metal pipe into the mouth so that the goat won't bite the tubing.
3. Thread the tubing through the pipe (or directly into the mouth) and gently guide it down the esophagus. If something blocks the tube's progress, pull it out and start again.
4. When the tubing is in the stomach, gurgling sounds indicate you are in the right place. Coughing or air sounds may mean you have entered the lungs through the trachea; withdraw the tube and try again.
5. Attach a funnel or the outside part of a 60 cc syringe with an irrigation tip.
6. Hold the end of the tubing well above the goat's mouth to ensure good gravity flow.
7. Pour the liquid slowly into the funnel, keeping the tube straight. Gravity will pull the liquid into the stomach.
8. To ensure all the fluid has drained, wait several seconds after the last of the liquid has gone down the animal's throat.
9. Before removing the tube, cover the end with your finger to prevent liquid from leaking and entering the lungs. Remove the tube slowly.
10. If the animal is weak or unconscious, prop it up on its chest. Never lay a goat on its side after tubing or it could aspirate fluid.

Administer only small amounts of liquid at any one time. Some breeders dip the tip of the tube into warm water to soften it before insertion. Another hint: Pour a small amount of water into the tube at the end of feeding to rinse the tube before withdrawal. Water, rather than a foreign fluid, is easier for the goat's lungs to handle and less likely to cause pneumonia.

HUMANE SLAUGHTER

Goats quickly lose the will to live when they are in pain. When a goat is too deformed, ill, or injured to survive, the kindest option may be to euthanize. The word *euthanasia* literally means "good death" in Greek. Discuss euthanasia options with your veterinarian and create a plan as part of your emergency medical kit. A vet can be called to the farm to give a lethal injection. When a vet is not available for this service, the herd owner may need to perform the task.

As stressful and unpleasant as it may be, preventing unnecessary suffering of the animal should be paramount in any situation with sick goats. When performing euthanasia, follow these guidelines:

- Goat skulls are very strong in front, so a gunshot should be taken either from the back of the skull between the ears, with the muzzle of the gun pointing downward through the mouth and between the teeth, or just behind the ear, aiming toward the opposite eye.
- In the absence of a gun, a traditional ritual slaughter technique may be used by cutting the throat straight across both jugular veins with a very sharp knife. The rapid blood loss is effective but messy and disturbing.
- Suffocation or drowning are not recommended.
- Barbiturates and anesthetics are available only to veterinarians. It is unacceptable to use over-the-counter products as a substitute for veterinary chemicals.
- Confirm that the animal is dead by checking whether the heartbeat, respiration, and corneal reflex are all absent.

If you intend to perform euthanasia yourself, know your local laws in order to avoid prosecution under animal-cruelty statutes. What is standard practice in one part of the country may be considered cruelty in another, so do not assume. A number of years ago, a prominent dairy-goat owner was charged with felony mistreatment of animals for putting down terminal kid goats using a blow to the head. While those charges were later dropped, the adverse publicity and $5,000 in legal fees were extremely damaging.

FINDING A VETERINARIAN

One of the goat owner's biggest challenges can be finding a competent veterinarian. Goats are not covered to a large degree in veterinary school. To find a good caprine vet, you'll need to talk to local goat owners and possibly interview a number of vets.

I consider myself lucky. I live in an area with a number of highly knowledgeable and skilled practitioners. Still, I called a number of vets before settling on my current one. One veterinarian, who shall remain nameless, gave me a lecture about why he won't treat goats. He really wanted me to know how much of a pain hobby-farm goat owners are!

To gauge the vet's overall skill, watch how the doctor handles the animal: Is care taken to be gentle? Willing to discuss treatment options with you? Knowledgeable about basic goat vaccinations and diseases? Willing to admit lack of knowledge and look up the information when unsure?

If you see something that bothers you, question the vet. Do not hesitate to request that the vet either change the treatment practice or leave your farm. The response will enlighten you as to whether you want to use the vet again.

Sometimes an incident or a mishap will leave no doubt. One farmer quit using a vet after the doctor threw the contaminated contents of an abscess directly onto the barn floor after removing it from a goat. And I suspect that a vet infected my first two goats with CAE during a routine blood test. He used a needle that had been used previously on other goats at the farm where I took them to be tested.

In another alarming case, a vet performing a difficult delivery finally managed to extricate the kid. The vet tossed the kid against the wall of the barn, stating, "That one is dead!" The horrified owner retrieved the kid, swung it gently by the hind legs, and revived it. She never used that vet again!

On the other hand, do not judge a vet by things that go wrong in the normal course of farm life. One of our goats died in elective surgery to remove an abscess located close to the jugular vein. A bleed occurred and two vets worked very hard to save her, but she died anyway. They did their best in a bad situation.

Your veterinarian is your resource for prescription medications, off-label drug use, and a myriad of medical skills that the basic goat owner can't begin to learn. Labels can also quickly become out-of-date, since medical advances happen so quickly. A good veterinarian has the resources to find current information.

Be aware that it is against the law to use off-label drugs without the advice of your veterinarian. The Animal Medicinal Drug Use Clarification Act of 1994 allows vets to use certain drugs "in a manner that is not in accordance with the approved label directions" if the owner and vet have a client/patient relationship.

CHRONIC CONDITIONS

Chronic illnesses may start with an acute condition or may lie dormant and never show in an individual goat. The disease may be infective to other animals. Some of the most troublesome current diseases are CAE, CL, Johnes, and chronic wasting diseases such as scrapie.

Conditions including brucellosis, tuberculosis, and bluetongue may be tested for depending on where you live and where your goat is going if you're preparing to move it. Some buyers request testing prior to buying a goat. Diseases such as tuberculosis in the United States and hoof-and-mouth in Great Britain must be reported to health agencies. Often, animals with reported infectious diseases are subject to quarantine and disposal.

Some genetics are known to protect goats from certain diseases. A test is being developed to test for scrapie resistance, following the success of a genetic test used in sheep. A genetic disorder currently being studied in Nubian goats is mucopolysaccharidosis IIID, or G-6-Sulfase deficiency (G-6-S). As this disorder becomes better understood, Nubian breeders and buyers are testing for G-6-S with increased frequency.

CLOSED HERD

The strictest form of disease management involves testing and culling any positive animals via slaughter. Never sell a goat that has tested positive for a disease unless you fully disclose the animal's condition to the buyer. The goal of this management style is disease eradication.

To ensure no diseases are brought to the farm, a strict closed herd does not buy animals, show animals,

or allow other goats onto the premises for any reason. Some herds modify these restrictions to allow showing and the purchase or breeding of "test negative" animals. Most closed herds test all animals on a regular schedule to catch any false negatives from previous testing.

Buying an animal from a closed herd maximizes your chances of getting healthy stock. Not surprisingly, these goats are normally more costly.

ISOLATION

Even serious diseases such as CAE and CL do not have to be a death sentence for the goat. Many goats have lived long, productive lives after receiving positive test results. Isolation achieves the ultimate goal of eradicating the disease in the herd while recognizing the difficulty of putting down asymptomatic, well-loved, or valuable animals.

Some owners keep two separate herds: one with positives and one with negatives. All "test positive" animals are handled last, often with different equipment, to prevent inadvertent spread of the disease. All kids, or sometimes just those from positive does, are bottle-reared on pasteurized milk.

COMINGLING

Some goat owners choose to let "test positive" and "test negative" goats comingle in the same herd. If you have a small herd or less time or space to devote to your herd, this may be the best option. Of course, goats that have visible symptoms or open abscesses should be removed from the herd by culling or kept isolated until open wounds are healed to prevent further spread of the disease. Often, the goal of this type of management is to raise disease-resistant, hardy animals with strong immune systems.

COMMON CAPRINE DISEASES AND AILMENTS

ABORTIONS

Toxoplasmosis, chlamydiosis, vibriosis, and brucellosis are infectious diseases that can cause spontaneous abortions. Pesticides, wormers, and other medication may also cause abortions.

Toxoplasmosis is a disease that is transmissible to goats, other animals, and even humans. It is carried by barn cats through their waste. Pregnant does who contract "toxo" may abort or have deformed kids. Keeping cats out of feeders, neutering them, and keeping only mature felines on the farm will minimize this risk. *Jen Brown*

One or two spontaneous abortions a year in a large herd is not usually cause for panic. An "abortion storm" is a situation when several, or sometimes most, of the does in a herd spontaneously abort in the same season. This is heartbreaking.

Care should be taken when working with delivery fluids, fetuses, or afterbirth. Toxoplasmosis and chlamydiosis are transmissible to humans. Because abortion has so many possible causes, a veterinarian should be consulted.

BLUETONGUE

Bluetongue viruses in the United States have long prompted officials to block the export of American goats (as well as cattle and sheep) to many world markets such as Australia, New Zealand, and Europe.

Transmitted by insects, bluetongue in goats is mild. Signs include inflammation and bleeding of the mucus membranes of the mouth, nose, and tongue, plus soreness and swelling of the feet. The condition is called bluetongue because of the color of the tongue or mucus membranes.

BROKEN BONES, SPRAINS, AND STRAINS

Goats have accidents, and breaks or sprains are not uncommon. Sometimes, goats twist a foot or a leg in a hole in the ground or catch a limb in a fence panel or feeder. Jumping off objects or being rammed by another goat can also cause injury. Examine any goat that is limping or favoring a leg.

A strain or sprain may be treated with liniment rubs and anti-inflammatory medications such as aspirin, willow-bark, phenylbutazone, or Banamine.

Unlike horses, goats can recover nicely from broken bones. Your veterinarian will x-ray the goat and then recommend treatment.

BRUCELLOSIS

Brucellosis is an infectious disease caused by *Brucella* bacteria. Sheep, goats, cattle, pigs, and dogs—as well as wild animals like deer and elk—are vulnerable. Humans can become infected. Symptoms of infection in humans include flu-like fever, sweating, headaches, back pains, and physical weakness. Severe cases may become chronic.

Only a few hundred cases of brucellosis are reported in the United States each year. This low number is due to the strict testing of livestock. The U.S. Animal Health Association believes that the country is now free from *Brucella melitensis*, which infects goats. In spite of this recommendation, some shipping regulations require testing anyway.

CAPRINE ARTHRITIS ENCEPHALITIS

First diagnosed in goats in 1974, caprine arthritis encephalitis (CAE) is a retrovirus similar to HIV in humans and feline leukemia in cats. CAE lives in the white blood cells and passes through body fluids. CAE is only transmissible to goats. It is most common in dairy

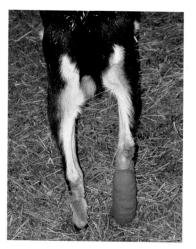

Broken limbs can be splinted and wrapped in Vet Wrap, allowing the goat to heal. *Carol Amundson*

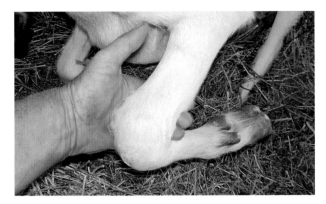

Arthritis can be crippling, especially when caused by CAE. This goat can no longer extend her leg. Often, the first sign of CAE is swollen knees. *Jen Brown*

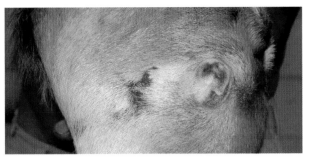

The lump, scars, and healing wound on this goat's throat are likely to be CL, but only a culture can confirm the diagnosis. *Jen Brown*

goats because of the practice of feeding pooled milk. There is no known treatment.

Goats infected with CAE may never show signs of illness. Rarely, kids younger than six months develop a fatal encephalitic form of the disease. CAE's primary visible manifestations appear in older goats as swollen knees, leading progressively to a crippling arthritis. Lowered immune capabilities, pneumonia, congested udders, and lowered milk production have been linked to this virus.

CAE Prevention is a management practice involving the removal of newborn kids from the dams and raising them separately from the herd. Birthing fluids and milk are the most common routes for spreading CAE. Feeding kids heat-treated colostrum and pasteurized milk or milk replacer prevents them from becoming infected through their mother's milk.

Long-term contact between healthy goats and latent carriers may spread CAE. Using the same needle to inject multiple goats can spread the virus. The same may be true if bloodsucking insects, such as ticks or lice, move from goat to goat.

The retrovirus lives only a very short time outside of living cells. It is killed in the environment by drying or by coming into contact with bleach.

CASEOUS LYMPHADENITIS

Abscesses, lumps, and bumps can be signs of caseous lymphadenitis (CL), a disease of the lymph nodes. CL results in tremendous cost to commercial herds from the

Lumps on the neck are a disqualification at most goat shows, as they can indicate CL. Be sure not to administer vaccines in the neck, so the vaccination lumps don't become confused with CL lumps. *Jen Brown*

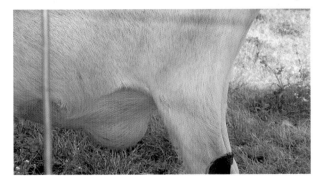

The lump on the doe's abdomen is from an abdominal hernia and is not infectious. *Jen Brown*

loss of revenue from condemned carcasses, the inability to show goats that have visible lumps, and the loss of sales to owners who want CL-free herds.

The organism that causes this disease is highly resistant, and no real treatment exists. Lumps or swollen glands in the area of the lymph nodes are the most common symptoms. Internal abscesses may form in the lungs or other organs. The CL organism may cause infection in people, too. Pus from broken abscesses infects the environment. Other goats can pick up CL from rubbing against common fences, feeders, and other contaminated surfaces. In a show setting, pens that have previously held animals with active CL could infect the goats placed in that pen.

An abscess progresses from a small lump to a ripe capsule that has little or no hair cover. These ripened abscesses can be cleaned out and flushed with diluted Nolvasan or other disinfectants. Not every abscess indicates CL, but only a culture can tell for certain.

Goat keepers that have chronic CL problems in their herd may decide to vaccinate. Breeders need to remember that vaccinated goats will likely test positive for CL because their body has received a small amount of the infectious organism.

DIARRHEA

Let's talk poop! I can't stress enough: Watch your goats and note changes in behavior, physical condition, and what goes into or comes out of their bodies. Normal goat droppings are round, firm pellets referred to as "nanny berries." The softer or looser the stool, the greater the chance the goat is sick. Minor dietary changes can cause log-style droppings, a pudding consistency, or minor diarrhea.

Some diarrhea is watery and squirts from the sick goat with every cough or movement. This is called scours. It means your animal is very sick and can become dehydrated and possibly die very quickly. To determine the treatment, first determine the cause:

- **Review any changes to your goat's diet or feeding environment**. Have you changed the type or amount of feed? Has the animal been grazing in a new or different pasture that has a different or more abundant vegetation? Taking your ailing goat off rich grain and switching to just hay will help.
- **Consider the age of the animal**. Is it a newborn or just a few weeks old? Scours is a common caprine

Normal nanny berries are round, firm pellets *(above, left)*. Pudding-like poop is abnormal and a sign that your goat is likely sick *(above, right)*. *Carol Amundson*

problem during the first few days of life. It may be caused by chilling, erratic feeding, dirty bedding, dirty milk bottles, or overeating. Spectam Scour-Halt (a pig medication) does a good job on our very young kids that have diarrhea. Warning: *Do not* use this medication on adult goats, as it works so well that it will destroy the working rumen's natural organisms and shut down the goat's digestive system. Scours in a three- to four-week-old kid is likely due to coccidiosis.

- **Consider the color of the diarrhea**. Is it green or black? Green diarrhea may indicate an eating change or plant poisoning; black, tarry stool is a sign of bleeding in the digestive tract—possibly due to coccidia or parasites. Yellow or light brown diarrhea tends to be from bacterial causes.
- **Check for internal parasites**. Worms can cause scours. Check the stool for the presence of parasites. This can be done by your veterinarian, or you can do it yourself if you have a microscope and reference materials. Proper deworming should take place if parasites are discovered.
- **Enterotoxemia type C**, also known as "bloody scours," is caused by the clostridium bacteria and usually seen in the first few weeks of life. The disease causes bloody infection in the small intestine.
- **Enterotoxemia type D**, also known as "pulpy kidney disease," is another bacterial disease, caused by a different strain of clostridium, that affects kids over four weeks old. It often affects the fastest growing, biggest kids.

Causes of Diarrhea in Goats

Bacterial	*E. coli* *Salmonella sp.* *Clostridium perfringens*
Parasites	Stomach worms
Protozoa	Cryptosporidium Coccidia (*Eimera sp.*) *Giardia sp.*
Diet	Dirty water buckets or those fouled with stool Feed changes or too much feed or indigestion Poor-quality colostrum or not enough within the first 24 hours Poor-quality or improperly mixed milk replacer Rich, wet pasture—especially sudden exposure Roughage—too little or not enough dry matter Toxins or allergies
Management (Poor environment)	Overstocking/overcrowding Poor sanitation
Stress	Changes or extremes in weather Changes in routine Traveling or shipping Weaning
Viral	Rotavirus Coronavirus

Both illnesses can start due to sudden feed changes. Affected kids are usually eating a large amount of grain or nursing does that produce lots of milk. Adults can also become sick with enterotoxemia after sudden feed or weather changes. Clostridial organisms are common in the environment and in the guts of ruminants. When feed change disrupts the balance of gut flora, the wrong organisms may take over, creating toxins that can cause indigestion or even death. When a herd becomes infected, losses may be high.

ENTROPION

Sometimes a newborn goat's eyelid turns inward, causing the eyelashes to rub against the eye and irritate the cornea. The kid's eye waters, crusts over, or swells shut. When caught early, entropion is easily repaired.

Injecting a small amount of penicillin directly into the eyelid causes the lid to swell and pulls the eyelashes away from the eye. In a few days, the swelling goes down and the eye becomes normal. You can also use a small clip or stitch the lid back after rolling the lashes outward.

FLOPPY KID SYNDROME

Floppy kid syndrome is another problem caused by clostridial organisms. Usually seen in kids under three weeks of age, symptoms include weakness, wobbly walk, and poor muscle tone—hence the name "floppy." When gently shaken, the kid's stomach will slosh.

Treat kids that are still active for three days with oral doses of penicillin, fortified vitamins, and thiamin. Pepto-Bismol or baking soda is helpful when the kid's stomach is full and nonfunctioning. Kids that are completely floppy may need to be tube-fed and given an injection rather than an oral medication. Do not tube milk into the stomach of flat kids, but instead use electrolyte and dextrose solutions until the kid has been up and moving for at least eight hours.

FOOT ROT

Foot rot and foot scald are hoof infections seen in warm, wet weather. These problems are more common in sheep than in goats. Affected animals may start out with a mild limp. Left untreated, the disease spreads to all four feet and makes the goat unwilling to walk or eat. Muddy confinement areas, crowded conditions, and rainy weather facilitate this disease. Affected animals should be treated by trimming the hoof and using a foot bath of copper, zinc, or iodine.

FROTHY BLOAT

Frothy bloat is a potentially lethal condition brought on by binge eating. Closely monitor any goat that has binged on the family garden or gotten into the feed room. Symptoms include a distended abdomen, piteous crying and moaning, and obvious discomfort. Call your vet, but also start treatment at home immediately. To break up the froth, drench with 1 to 2 cups of vegetable or mineral oil or a commercial bloat medication. To break up the gas, gently roll an adult goat back and forth on the ground or jiggle a kid up and down on your lap.

GLUCOSAMINE-6-SULFATASE DEFICIENCY

Glucosamine-6-sulfatase (G6S) deficiency is a recessive genetic disorder prevalent in Nubians. One out of four are G6S carriers. Carrier animals are healthy, but kids that inherit G6S genes from both parents have the disorder. G6S-deficient goats lack an enzyme needed for growth. Animals may die young, or they may live as long as four years in weakened condition—allowing them to breed and pass along the genetic defect.

Nubian breeders can avoid the disorder by knowing the genetic status of at least one parent and breeding appropriately.

HOOF-AND-MOUTH DISEASE

An extremely contagious viral disease seen in cattle, pigs, goats, and other livestock is hoof-and-mouth disease, also known as foot-and-mouth disease. Humans can catch it in very rare cases. Affected livestock have blisters and sores around the lips, the tongue, the teats, or the coronary band of the hoof. Goats become lame and sometimes drool. They also have difficulty eating when the sores in the mouth are numerous. Death may occur.

There have been no hoof-and-mouth cases in the United States since 1929 and none in Canada or Mexico since 1954. Currently, animal import bans from affected parts of the world keep this disease out of the United States. However, the extremely virulent nature of hoof-and-mouth was seen by the destruction of many cattle in affected parts of England in recent years. Vigilance is important.

JOHNES DISEASE

Johnes disease is a fatal chronic wasting disease of livestock similar to tuberculosis and leprosy in humans. Affected goats are not always symptomatic, making it hard to diagnose. Goats showing active Johnes hungrily consume feed but look thin and scruffy. While diarrhea is common in cows with Johnes, the symptom is seen in goats only at the very end or not at all. Once present in a herd, it is hard to get rid of Johnes. Highly contagious, it is spread by actions as simple as a kid nibbling infected hay. This infection is not treatable. The best way to prevent Johnes is to use CAE prevention techniques— raising kids separate from the herd.

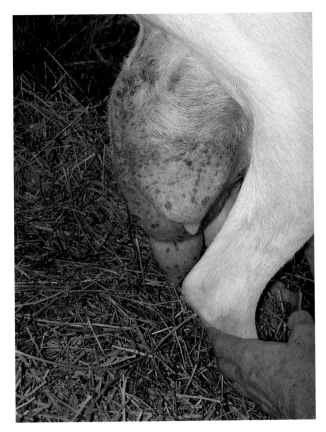

Infections of the udder, called mastitis, can cause deformity
and ruin half or all of the milker's productive ability.
Jen Brown

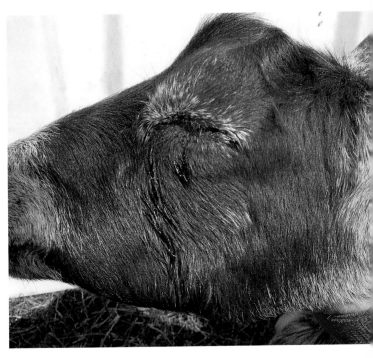

Tearing, cloudy eyes, and blindness are signs of pinkeye.
Jen Brown

MASTITIS

Mastitis is inflammation of the udder caused by
infection, seen most commonly in the dairy barn and in
nursing dams. Symptoms include a hot or swollen udder
and milk that has lumps, strings, thickening, bad odor or
taste, or an off color. Teat infusions, either medicated or
holistic, are used to treat this condition. Some breeders
report success with high doses of vitamin C administered
orally or infused. A dry treatment between lactations
will help clear up subclinical mastitis.

NEUROLOGICAL PROBLEMS

Neurological problems are caused by deer worm, listeriosis, polioencephalomalacia (PEM), and any other
disease that causes inflammation of the brain. Signs of
neurological impairment include staggering, circling,
and gazing off into the distance or straight up at the sky
("stargazing"). Extreme cases suffer from convulsions or

blindness. Treat with massive doses of fortified B vitamins several times a day—and call your vet. Depending
on the cause, antibiotics or wormers may be necessary.

PINKEYE OR INFECTIOUS
KERATOCONJUNCTIVITIS

Eye ailments are frequently seen in the summer, when flies
spread the problem between goats. Pinkeye causes red,
weeping eyes. Severe cases progress to cloudiness and,
eventually, blindness. Keratoconjunctivitis is inflammation due to dryness. Both diseases are highly contagious.
Do not use the cattle vaccine for caprine pinkeye. Treat
mild cases with medicated eye-drops, tetracycline ointment, or a puffer. Use injectable oxy-tetracycline in more
severe cases.

RESPIRATORY ILLNESS AND PNEUMONIA

Respiratory illnesses are caused by both bacteria and
viruses that may initially lie dormant in goats and then
be triggered by a stressor. Sometimes called "shipping
fever," caprine pneumonia is frequently seen in animals
moved from one location to another. Changes in weather
may also trigger an outbreak.

Vet RX nose-drops are a non-medicated method of relieving minor sniffles. *David Weber*

Nasal discharge, coughing, fever, labored breathing or rattling, and depression signal serious respiratory problems. Consult your veterinarian for treatment. Mild cases can be relieved using Vet RX, Vicks VapoRub, or other herbal remedies that contain menthol or eucalyptus.

There is a pneumonia vaccine for cattle, but its use in goats is controversial.

RINGWORM

Another disease frequently brought home from goat shows is ringworm, a painful nuisance to the goat. It can be transmitted to humans and other pets. Goats lose hair in scaly, circular patches. Antifungal shampoo or athlete's foot spray are simple, over-the-counter treatments.

SCRAPIE

Transmissible spongiform encephalopathies (TSEs) are of concern for both the economic loss of sick livestock and the potential for human infection. These conditions are known as mad cow disease in cattle, chronic wasting disease in deer, and scrapie in sheep and goats. Fewer than twenty cases of scrapie have been reported in goats in the United States since 1990.

This is a slowly developing neurological disease. Destruction of suspected animals and postmortem examinations are the only way to confirm the disease. If a breeder who keeps both sheep and goats has a diagnosis of scrapie in the sheep herd, all of the sexually intact goats must be destroyed. Breeders with valuable goats might want to consider not keeping sheep or, alternatively, keeping sheep that are genetically resistant to scrapie.

Scrapie eradication programs require exhibited animals to be from farms belonging to a scrapie ID program.

SORE MOUTH

Sore mouth, also known as orf or contagious ecthyma, is a common but painful condition characterized by blisters, pustules, and open sores on the mouth of a goat. It can also affect the teats, ears, and other places that a kid has tried to nurse. The biggest concerns caused by sore mouth are the failure of kids to eat, secondary infection of the sores, and contamination of the environment. Scabs that fall on the ground can infect the environment for years to come. Sore mouth is transmissible to humans.

Livestock shows prohibit animals with sore mouth. Unfortunately, within two to ten days of returning home from an ostensibly clean show, newly infected animals will start showing signs of disease. As a preventative measure, isolate animals that are new to the farm or returning from shows.

Mostly an annoyance, sore mouth generally goes away on its own within six weeks. Udder balm on infected teats can help reduce pain. Kids that are not eating may need to be tube fed. Antibiotic treatment for secondary infection may be necessary in severe cases.

A vaccine made with live organisms is available. Use it only if your herd has had the disease; otherwise you're likely to bring it onto the premises by the very means intended to prevent it.

TETANUS

Tetanus is an often fatal disease caused by a bacterial toxin. Animals and people contract tetanus through a deep wound or cut that doesn't bleed properly. Castration and disbudding may cause infection from tetanus. Also called lockjaw, symptoms include spasms, loss of coordination, and stiffness, especially in the jaw.

Goats with sore mouth may go off their feed, leading to secondary problems. It is highly contagious. *Jen Brown*

Prevention is always better than treatment, and this is especially true for tetanus. It is strongly recommended that you vaccinate your goats with CD/T, which gives immunity to Enterotoxemia C and D and tetanus. In young kids from unvaccinated mothers, give the tetanus antitoxin for short-term protection.

The prognosis is not good for a goat that has contracted tetanus and started to stiffen. Treat with large doses of penicillin and tetanus antitoxin. Holistic practitioners believe that vitamin C detoxifies tetanus victims.

TUBERCULOSIS

Tuberculosis has not been found in goats in the United States for many years, but it continues to be monitored due to its prevalence in cattle. A simple skin test can be done by a vet to see if a goat has had contact with the disease. This skin test is often required for interstate shipping or export.

URINARY CALCULI

Urinary calculi are tiny stones or crystals that form in a goat's urinary tract, causing pain and blockage. Bucks and wethers are most susceptible to this condition. Symptoms include straining to urinate, dribbling, blood in the urine, or pain exhibited by grinding teeth or lying down and kicking at the abdomen. Many cat owners are familiar with this problem. A blocked urethra is life-threatening. Call a vet if blockage is suspected. Prevent calculi by providing plenty of fresh drinking water and supplementing the diet with ammonium chloride.

TRAVELING WITH YOUR GOATS

Above and opposite: **Goats often love to go traveling once familiar with the vehicle. Our goats gather around the trailer because they enjoy traveling to shows.** *Jen Brown*

If the owner of the goat is not afraid to travel by night, the owner of a hyena certainly will not be.
—Nigerian proverb

Most goats are excellent travelers. Because of their small size, goats can be transported in many different vehicles. The key factors are the same as in housing: Avoid wind, extreme temperatures, and wet conditions. In trailers or carriers, the sides should be smooth with no protrusions or sharp edges. No carrier should be totally enclosed because poor ventilation can cause stress and possibly suffocation, especially in hot weather. Exhaust fumes from other vehicles seeping into poorly ventilated spaces can cause carbon monoxide poisoning.

Use common sense when moving your animals. Those that are sick, lame, or heavily pregnant need special care. These goats should be transported only under conditions of need, such as veterinary care. Tying a goat in the back of an open vehicle could lead to injury or strangling. Tying a ruminant so that it's lying down on its side can cause bloat.

Standard transport for goats is either the bed of a covered pickup or a trailer. Use bedding such as straw unless you are going a short distance. Bedding provides a comfortable surface and soaks up animal waste. Slippery floors can cause goats to fall.

When loading a truck or trailer, make sure there is enough room for each animal to move comfortably. The FAO recommends 4.5 square feet per goat. In an overloaded trailer, a fallen goat can be trampled. When driving a truck or trailer with animals in the back, drive smoothly, avoiding jerks or sudden stops. Make turns gently.

Feed and water your goats before a long trip to keep them more comfortable. Goats need feed and water every twenty-four hours at minimum on long journeys, and it is best if they can be offloaded to stretch their legs. This rule is applied to meat animals being shipped for slaughter but also pertains to goats moving cross-country for shows or to a new home.

Mixing horned and hornless animals is risky. The same is true for mixing kids and adults or bucks and does. It helps if the animals are familiar with one another so that they won't have to to settle questions of rank in the confines of a truck. Naturally, a randy buck or several bucks with does is a recipe for disaster. I mix adults and kids only when the truck is partially full and the kids are traveling with their dam.

Small-breed goats or kids fit nicely in travel kennels used for dogs. The larger goat breeds can travel in similar boxes that have been specially made to fit in a truck or car. There are even cages designed to go in the bed of a pickup for short rides in nice weather.

Just like dogs, goats will ride comfortably on the back seat. It is wise to have a mat or blanket and a well-mannered, tame animal when trying this method of transport.

INTERSTATE SHIPPING

A myriad of rules cover the transport of animals across state lines. Usually, a minimum requirement is an interstate Certificate of Veterinary Inspection. Show rules usually spell out the conditions for out-of-state exhibitors. Find out what requirements are needed by your destination state and any states you are traveling across. A good resource is the USDA Animal and Plant Health Inspection Service. Its website (see Appendix) gives links for individual state and international regulations.

FLYING A GOAT

Most goats that travel by plane are kids purchased for replacement or breeding stock. A small goat breed or a kid can often be shipped as a "pet." Different airlines have different rules, but all are governed by Federal Aviation Administration and USDA standards. Some basics guidelines: Provide a standard dog crate large enough for the goat to stand up and turn around. Attach food and water to the crate. Provide a collar and leash for ease of inspection. When shipping goats in cold weather, provide a veterinary acclimation statement saying that the animal can withstand the temperature extreme. Do not ship goats in hot weather. Be prepared for the inevitable airline searches, delays, and mix-ups.

Above and opposite, upper left: A homemade wooden box for the bed of a pickup can be all you need for transport. *Jen Brown*

INTERNATIONAL EXPORT

Export rules have gotten particularly strict in light of biosecurity concerns. A number of borders have been closed to goats from the United States, including all of Europe. Other countries have explicit testing or quarantine regulations. Most owners who sell goats into another country work through a buyer who knows the needs of the customer. In some cases, the intermediary purchases the goat and performs the testing and quarantining.

RECORD KEEPING

Various regulations, as well as the proposed National Animal Identification System, require breeders to track the movement of their livestock. Keep records of the tag or ID numbers of any animals permanently leaving your property, the dates they left, and their destination. Similar records should be maintained for purchased goats.

With a nice blanket, the back seat is comfortable! *Jen Brown*

Getting into a pickup can be a tall jump for the goat. Some goats need a little help getting into the truck. We have others that jump right in. *Jen Brown*

KEEPING MILK GOATS

I find among writers that the milk of the goat is next in estimation to that of women, for that it helpeth the stomach.

—William Harrison,
Elizabethan England (1577)

The dairy products of goats—milk, cheese, yogurt, sour cream—are a healthy, tasty alternative to cow products. People accustomed to commercial cow milk will find goat milk richer, slightly sweeter, and varying in flavor with the season. Commercial cow herds contain animals in various stages of lactation. By the time their milk is shipped, processed, and packaged, its butterfat percentage has been adjusted and the flavors standardized. Fresh goat milk is as comparable to this product as fresh vine-ripened garden tomatoes are to the tasteless ones from the supermarket.

I love giving two glasses of milk to someone unfamiliar with this tasty product. One glass has whole cow milk, the other goat milk. When asked, "Which is the cow milk?" the person holds up the goat-milk glass. Why? "Because it tastes better."

As a health food, goat milk has a lot going for it. Although higher in fat than cow milk, the fats contain more omega-3 fatty acids, the same ones in fish oils touted for increasing "good" cholesterol. The fat molecules are also smaller, so they stay suspended longer than fat molecules from cow milk. This means that goat milk is naturally homogenized and only separates very slowly. You won't see a lot of homemakers using goat milk butter or whipping cream for this reason.

Many people become interested in raising goats because they have allergies or other health issues that prevent them from drinking commercial cow milk. They search out a local farm to provide goat milk on a regular basis, and soon they want the milk source at hand. That's when they buy a couple of dairy goats—and the fun starts.

HOME VERSUS COMMERCIAL DAIRYING

Home dairies operate at the whim of the owner. As long as the milking parlor is clean and the milk handling correct, no other rules apply. The goat owner who chooses to go commercial has an entirely different operation. Any commercial dairy must follow the regulations of its home state. Each dairy is licensed by its state and pays an inspection fee once or more each year. Interstate shipping regulations require additional inspections, licensing, and fees. Inspections cover everything from the milk holding temperature to feed storage to veterinary treatment.

Artisan chèvre cheese is made from fresh goat's milk.
Shutterstock

In a blind taste test, people unfamiliar with goat milk often find it tastier than cow milk. *Shutterstock*

MILK HANDLING

Milk is a wonderful food for goats and humans—and for microorganisms. It is impossible to produce completely organism-free milk, but microorganisms must be controlled. Good microorganisms help make yogurt and cheese and also make milk more digestible. Opportunistic microorganisms, though, can spoil the milk at the very least and cause illness in a worst-case scenario, such as infection with *E. coli* or staphylococcus.

In order to keep milk from spoiling, everything that touches the milk must be sterile. In the simplest terms, clean milk directly from the teat is filtered into a sterilized container. The first squirts of milk should be collected in a strip cup. Throw away the milk in the cup because

it could contain dirt and contaminants from the end of the teats. Sterile procedures must be followed whenever milk is transferred to other holding containers.

Ideally, a milk pail should be stainless steel with a lid. For commercial dairies, stainless steel, food-grade plastic, or clear glass are the only non-filter materials allowed to contact the milk. For private use, any container that can be bleached is fine.

Bleach is the best sanitizer for dairy equipment. Once the container or other implement has been washed, use a hot-water rinse of 2 percent chlorine sanitizer or bleach. Submerse the item in the sanitizer bath, then hang it or set it on a clean rack to drip dry.

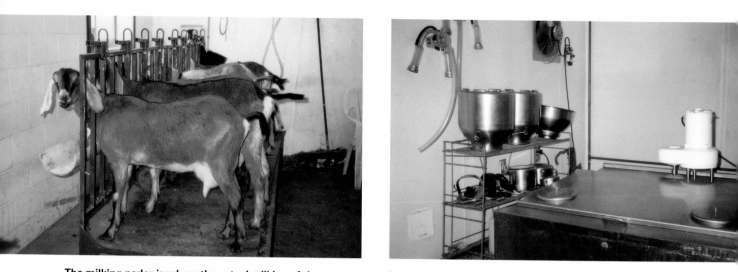

The milking parlor is where the actual milking of the goats occurs *(above, left).* **The processing of the milk occurs in a separate room called the milk house** *(above, right). Terrapin Acres*

Around 1862, the French chemist Louis Pasteur discovered that heating liquid killed germs. This discovery revolutionized food handling and safety. The spread of tuberculosis and brucellosis through contaminated milk has since been vastly reduced. For proper pasteurization, milk should be heated to 161 degrees Fahrenheit for fifteen seconds. Heated milk should be cooled rapidly to 36 degrees in an ice or cold-water bath and placed in refrigerated storage.

Raw milk is controversial. While I enjoy drinking our milk raw, I know the health status of my does and the handling of my milk. The American Veterinary Medical Association stands firmly on the side of pasteurization. In many states, it is illegal to sell raw milk.

Consumption of raw milk should be an informed decision. Do not use raw milk for infants or people who have compromised immune systems. I pasteurized milk for our daughter until she was two years old as a precaution against illness.

OFF-FLAVORED MILK
Normally, fresh goat milk is sweet with no strong or "goaty" flavor. If your goat produces unpleasantly strong-flavored milk, check your handling process. The longer milk is held in storage, the more the fatty acids break down, producing a stronger flavor.

The food your goat eats also affects her milk. Plants like ragweed or goldenrod may flavor the milk. Strong kitchen scraps can have an effect, particularly onions, garlic, cabbage, broccoli, grape leaves, or blackberries. The same principle applies if she is deficient in cobalt, vitamin B12, vitamin E, or protein.

Mastitis
In some cases, off-flavor milk is due to an infection of the udder known as mastitis. The milk may taste a little salty, or it may develop an off-color, stringy texture and bad smell. Lumps or strings can be detected by squirting the first streams of milk from each teat into a screened strip cup. During late lactation, mastitis is commonly subclinical, meaning that it's not noticeable except through testing. Commercial dairies regularly test milk for the organisms that cause mastitis. Blood in the milk may indicate infection. However, in early lactation, pink color is often due to the breaking of tiny blood vessels in the udder as it grows to accommodate the milk. Investigate any bluish color to the udder, excessive hot or cold feel to the skin, or sudden hardness or unevenness. Do not drink milk from infected goats.

THE DAIRY BARN
The dairy barn should be set up for convenience. Have water-access, milk-handling, and clean-up implements and supplies nearby for the home dairy. Commercial regulations specify everything from the way water is delivered into the area to what materials may be used in

the walls. In a formal milking parlor, the walls and floor are made of a material that can be washed. Flies must be controlled.

Milk stands are readily available in many styles. My first milk stand was homemade. Later, when I had my commercial dairy, I had double metal stands to hold two goats. These stands can be placed side by side, creating a row of milkers, yet they are easily moved for cleaning. A portable milk stand can be taken to shows. Concrete stands work well in a formal parlor with metal head gates built in. The concrete is easily swept or washed.

Commonly, each milk stand has an attached feeding trough or bin. This setup ensures that the goat is getting her feed and keeps her busy while you work.

Plan how the goats will come into the parlor. If you have only a couple of goats, they can just jump up on a stand. As number of goats increases, however, you'll need an orderly way for the goats to enter and exit. Some dairies accomplish this with a pen system that lets goats enter by one door and exit by another.

Additional equipment setup will depend on whether you milk by hand or by machine. Each procedure has its advantages and disadvantages. Hand milking requires low-cost equipment and is a peaceful way to spend time twice a day. Using your hands can also help you monitor udder health. Many goat owners refuse to milk by machine simply because of the mastitis risks involved with machine milking.

Machine milking has a high initial setup cost and requires technical knowledge. If the equipment is set up and used properly, however, machine milking allows for less experienced farmhands to help with chores. While some machines can be loud, locating the compressor outside the parlor eliminates much of the noise. Once goats become used to the machine, most will allow even strangers to milk them without complaint.

Milk handling and processing are usually completed in a room separate from the milking parlor. The milk house usually has water hook-up and cold storage. For the small home dairy, these steps are readily handled in the kitchen. Having a separate room close to the milking parlor for processing milk, especially for feeding milk back to kids, keeps the clutter of utensils and the mess of spilled milk away from the house.

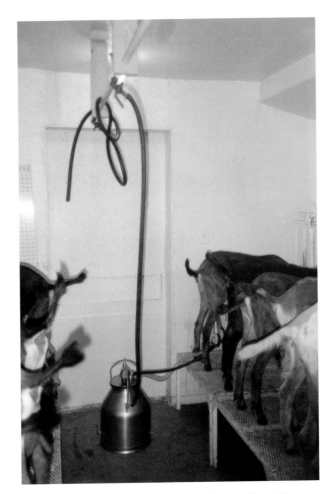

A formal milking parlor has metal head gates, elevated stands, and washable floors and walls. *Terrapin Acres*

Pigs love goat milk. Raising other livestock on goat milk can be fun and makes use of excess milk on noncommercial dairy farms. *Terrapin Acres*

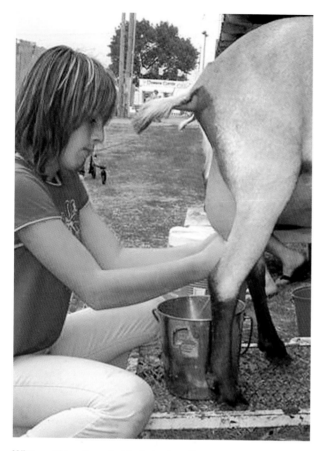

When milking from behind, the handler can sit without twisting—and see the entire udder at all times.
Carol Amundson

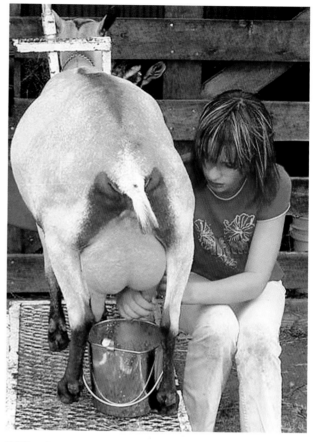

Milking from either side is more common. To prevent startling the goat, it helps to talk to her and let her know you are going to touch her. *Carol Amundson*

Hand-Milking Equipment

Milking stand
Stainless-steel milking pail
Stainless-steel strainer and milk filters
Dish soap or dairy soap
Bleach or sanitizer
Acid detergent

Clean-up brushes
Strip cup
Mastitis indicators
Paper towels or dairy towels
Teat dip such as Fight Bac®

The handler should grasp the udder gently but firmly. Allow the teat to fill with milk, then close off the top of the teat with your thumb and forefinger. *Carol Amundson*

HAND MILKING

By far the most common way to get milk from a goat is by hand milking. After a little practice, milking can become a great way to spend time with your goats. An adept milker can milk a goat out in three to five minutes. Do not be discouraged if it takes you longer at first. My first experience took twenty minutes, and I got more milk on me than I did in the pail—and the milk was brown because the goat's manure-covered feet got into the pail as well!

Goats can learn to be milked from either side or from the rear. Most often, milkers work from the side. I milk from behind because it is easier on my back, and I like being able to see the whole udder to check its condition. Unlike cows, goats usually don't kick backward and rarely urinate or defecate while being milked.

Squeeze the rest of the teat with the other fingers until the teat has emptied. Release the top of the teat and repeat the process until the udder is empty. *Carol Amundson*

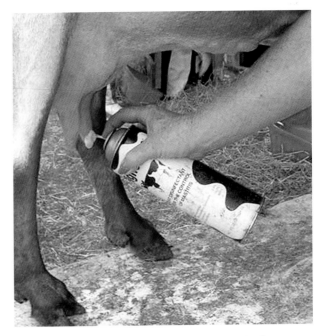

Spraying or dipping the teat with an udder wash after milking will help protect the doe from mastitis until the teat end closes off. *Carol Amundson*

KEEPING FIBER GOATS

Cashmere goats produce fine cashmere fiber for clothing and other products. *Shutterstock*

Behold, thou art fair, my love; behold, thou art fair; thou hast doves eyes within thy locks: thy hair is as a flock of goats, that appear from mount Gilead.

—Song of Solomon

Goats raised primarily for their hair coat are known as fiber goats. They produce cashmere or mohair fiber for handcrafts and spinning. The highest-value commercial fibers are normally white, because the uniformity of color is useful in mass manufacturing. Colored fleece is valued by hand-spinners and artisans who appreciate the natural variation in tone and hue.

The name Angora comes from the capital city of Ankara, Turkey, where these goats originated. Angora goats fill a commercial niche in the United States. Large commercial herds—predominantly in the Southwest—produced more than one million pounds of mohair in 2007. Census figures show that 185,000 animals were clipped, yielding an average of 6 pounds of fiber per goat. The mohair sold for an average of $3.78 per pound.

Cashmere goats, on the other hand, are not counted in census surveys. A Cashmere type fiber can be found on almost any breed, except Angora. In the early 1970s, researchers began investigating the undercoat produced by American goats. They found that many goats did produce a cashmere type undercoat but not enough to make it economically viable. Through selective breeding, we now have cashmere-producing goats in the United States.

Cashmere is warm, soft, and long-lasting. Because only 4 to 6 ounces of this fine undercoat are produced each year per goat, products made from cashmere are highly valuable. The price of cashmere products varies based on the quality of the fiber and the workmanship of the final product.

The Cashgora goat is a cross between Angora and Cashmere breeds. Angora crosses with miniature breeds are more likely to be found in hobby and pet homes rather than commercial herds. An Angora–Nigerian cross is called a Nigora, and an Angora–Pygmy cross is called a Pygora.

Angora goats produce mohair fiber. Pure white hair often commands the highest price. *Shutterstock*

Created breeds such as this Pygora goat are popular with artisans who like unique fibers. *Fran Bishop, Rainbow Spring Acres, Pygora Breeders Association*

BUYING FIBER GOATS

As noted in chapter 4, conformation should always be the first consideration when buying any goat. Look for strong legs and feet, a well-formed mammary system for raising kids, and a good set of teeth and jaw for maintaining proper nutrition.

After considering conformation, *next* look at the fleece. It should cover as much of the body as possible. If you are a knitter, spinner, or crafter already using fiber, you have an advantage in deciding what you want in your herd. If you're a newcomer to fiber animals, learn as much as you can before making the plunge. It costs as much to feed a poor-quality animal as a good one.

REGISTERING FIBER GOATS

Purebred Angora goats are registered with the American Angora Goat Breeders' Association (AAGBA), established in 1900. Purebreds must be white and have purebred parents. All Angora goats need two forms of identification, whether tattoos, ear tags, or ear notching. Ear notching entails placing notches and holes into specific parts of each ear as a code.

In 1999, another registry opened, the Colored Angora Goat Breeder's Association (CAGBA). If the animal has registered parents in either CABGA or AAGBA on both sides of the pedigree, the breeder may certify that the animal is free of disqualifying traits and register the goat without inspection. Blue Card goats are those with distinctly colored fleece, and Red Card goats have white or lightly colored fleece. Animals that do not have two registered parents must undergo inspection. Most often, this process occurs at a show or event that has inspections built into the program.

The various other fiber goats, such as Pygoras and Nigoras, have their own registries. In 2005, the Cashmere Goat Registry was established. This registry is a privately held registry, guided by a devotion to the improvement and preservation of cashmere producing goats, over bureaucracy and politics.

CARING FOR YOUR FIBER GOATS

The majority of fiber goats in the United States are raised on range conditions. This method of rearing presents challenges for meeting the nutrition and safety needs of the herd. Most of the literature about fiber goats addresses this style of rearing. With small flock owners, the emphasis shifts toward intensive management practices. Unfortunately, many people who keep these goats also keep sheep, and they often try to treat them as sheep.

Goats are more prone to chilling from cold or damp weather than sheep. Angora goats with 2 inches of hair growth can tolerate subzero temperatures. However, if you expose a herd of freshly sheared Angoras to sudden temperature drops or increased wind and humidity, the animals may become ill or even die.

Cashmere goats are normally hardier. In fact, the best-quality cashmere is said to come from goats living under rough conditions high up cold mountains. The same goats brought inside and pampered would no longer produce the same quality of fiber.

HARVESTING YOUR FIBER

CASHMERE

The cashmere undercoat grows throughout the fall and winter. The slight sheen on growing fleece is called "life." In the spring, as the daylight hours lengthen and goats approach kidding, the hair naturally begins to shed. The finest fibers are shed first, and the fleece becomes a dull matte in appearance.

Large commercial fiber operations tend to shear their Cashmere goats. This is done once a year in early spring, before the cashmere undercoat begins to shed. The fine cashmere fibers are then mechanically separated from the stiff guard hairs.

In small herds, the fiber is typically harvested by combing. Cashmere combs have long, sharpened tines with a bale that moves back as the comb becomes full of cashmere. A dog grooming rake works well, too. This traditional method mimics natural shedding by leaving hair on the animal and protecting against weather. Selective breeding for true cashmere and proper timing ensure that you can comb out the cashmere in one or two combings. If the guard hairs are longer than three times the length of the cashmere undercoat, trim them before combing.

Types of Fiber

CASHMERE

- Defined by the U.S. Wool Products Labeling Act of 1939 and the Cashmere and Camel Hair Manufacturers Institute as "the fine (dehaired) undercoat fibers produced by a Cashmere goat (*Capra hircus laniger*)."
- Fibers have an average diameter of 19 microns, with no more than 3 percent (by weight) over 30 microns in diameter.
- Protected by long coarse outer guard hairs that must be removed from the shorn fleece to preserve the quality of the cashmere.
- Adults produce 4 ounces of cashmere fiber per year.
- A single cashmere sweater requires the fiber of three Cashmere goats.

MOHAIR

- The long, silky hair of the Angora goat.
- Kid fiber is 4 inches long at shearing and finer than adult fiber.
- Yearlings produce 3 to 5 pounds of mohair.
- Adults produce 8 to 16 pounds of mohair a year.

CASHGORA

- The hair of large-breed (Cashgora) and small-breed (Pygora and Nigora) fiber crosses.
- Limited uses; typically not as fine as mohair or cashmere.

- Animals fall into three types:
  - Type A: Hair is similar to mohair and smooth and cool to the touch; requires shearing one or two times a year.
  - Type B: Hair is airy, light, fluffy, and soft and warm to the touch; harvested once a year by combing, plucking, or shearing.
  - Type C: Hair is creamy, suede-like, and warm to the touch; shows no luster; often commercially acceptable as cashmere in type; harvested once a year by combing, plucking, or shearing.

KEMP

- The long, straight, brittle, hollow hair that can show up on the thighs and backbone of any fiber goat.
- Fiber breaks easily and does not take dye well.
- The presence of kemp is a criterion for culling a goat from a fiber herd.

Mickey Nielsen, Northwest Cashmere Association

Mohair only comes from Angora goats, whereas cashmere is the soft undercoat produced by almost all the other goat breeds. *Shutterstock*

MOHAIR

Angora goats need to be sheared twice a year, usually before breeding and before kidding. Mohair grows about three-quarters of an inch per month on an adult goat and should be 4 to 6 inches long at shearing.

CONTAMINATION

The cleanliness of fleece is important. There are three kinds of contamination:

- **Natural:** Contaminants created by the goat itself, such as black or colored fiber, yolk (a combination of sweat and grease), urine, and dung.
- **Acquired:** Anything picked up from the environment, such as feed, vegetable matter, poly fibers from twine, even cigarette butts.
- **Applied:** Products intentionally applied to the goat, such as paint from brands, topical wormers, and fly spray.

To lessen fleece contamination, hold the goat a minimum of four hours without feed or water in a clean, dry pen before shearing. To shear a goat, start with a clean floor or put down a piece of plywood to catch the fleece. Sweep the shearing board clean after each shearing. Cashmere goats are sheared standing up with their head in a stanchion or head bale. Cut close to the skin the first time, as second cuts result in uneven fibers that lower the value of the fleece.

Each goat's fleece is placed in its own bag. Using a permanent marker, the bag is then marked with:

- Grower's name
- Fleece type (kid, yearling, young adult, adult, buck)
- Goat's name and identifying number
- Date
- Clip season (spring or fall)
- Problems (burrs, long fleece, short fleece)
- Fleece quality

An Angora kid has soft, clean white hair. The cleaner you keep the fleece, the better price you will get for your fiber. *Shutterstock*

FIBER MARKETS

Just like goat milk, goat fiber is a specialty commodity that has various outlets. There may be a local cooperative in your area that fiber producers can join that markets bulk fiber. Some fiber producers sell fleeces directly to handworkers for up to $12 per ounce; others process their own fiber and sell it as yarn or other products. Commercial facilities will process fiber for a fee, then the farm sells the cleaned mohair or cashmere. Only three fiber mills in the United States can successfully dehair fine cashmere.

SECONDARY MARKETS

Don't forget that when you raise a fiber herd, you also have a secondary meat market. Excess males and goats that do not produce quality fiber are a fiber industry byproduct. The Angora goat herds of the Southwest are a major source of meat goats. Many agricultural census tallies have cashmere goats counted in the meat goat category.

Cashmere is a highly sought product. *Shutterstock*

ENJOYING YOUR PET GOATS

Pet goats such as this Nigerian kid can be wonderful companion animals. *Jen Brown, Cutter Farms*

The goats have taught me a lot in the past thirty years. They don't, for example, care how I smell or how I look. They trust me and have faith in me, and this is more than I can say for a lot of people.

—Ches McCartney, the Goat Man

Goats are smart and personable. They pair the friendliness of dogs with the intelligence of cats. Inquisitive by nature, goats explore new surroundings, test fences and gates, try tasting everything, and generally poke their noses into whatever is within reach.

These qualities make goats wonderful pets—or real problem animals, depending on their upbringing. Just as an undisciplined dog causes problems for the owner, a goat becomes a nuisance when untrained.

RULE 1: TEACH BASIC COMMANDS

A goat can learn simple commands just like a dog. Choose simple words and use the word consistently to reinforce behavior. When training, accompany the com-
mands with direction from your hand or a squirt from a water bottle. Some form of physical contact should be used at first. Chasing the goat is counterproductive and may teach it a new game: running away! Some basic, useful commands include:

"**No**": Stop what you are doing.

"**Stop**" or "**Back**": Don't push and don't rush the gate.

"**Come**": Move toward me.

"**Hup**," "**Hey-up**," or "**Up**": Jump into a trailer or onto a milk stand.

RULE 2: NO PUSHING

Whether your goat has horns or not, allowing it to play by pushing with its head is a bad idea. Goats settle disputes between each other by butting heads. When you push back at the head, you teach the kid that this is an acceptable way to play with humans.

Never invite head pushing or butting. Do not grab the goat by its horns. When the goat lowers its head, hold

A well-disciplined goat is a happy goat. One basic command is "Hey-up," meaning jump up, in this case onto a hay wagon. *Jen Brown*

out your hand and issue the command "No!" or "Back!" Gently redirect the goat by placing your hand on the side of its face. If the goat continues to butt, accompany your command with a squirt from a water bottle or a sharp tap on the end of the nose. Never hit the goat, or it will learn to fear you.

RULE 3: NO JUMPING

Pet goats should not jump on people. To address this behavior, gently grab the goat by the shoulders and set its feet firmly on the ground. Accompany this physical contact with the command "No!"

For persistent bucks, it may be necessary to be more forceful. Try grabbing the goat when he is in midair and forcing his head and shoulders all the way to the ground. Touching his head to the ground teaches him that you are dominant.

RULE 4: CONTAIN THE GOAT

"We had a goat when I was a kid and he always climbed on cars."

"All of my roses are gone, and the lilacs have been chewed off to the ground."

"The goats got into my neighbor's trees and stripped bark off their trunks. Some of them were special ornamentals, so if they die, I have to replace them."

Any breed can make a good caprine pet. Nubians are beautiful and popular, but every breed has something to offer. *Jen Brown, Terrapin Acres*

Over the years, I have heard these stories from people visiting my farm to give up their goats. Unsupervised goats running loose in the yard will get into trouble. When supervised, the goats can be redirected from bad behavior. Alone, goats should be confined to an enclosure.

Tethering or tying a lone goat out on a lead is not a good idea. A goat staked out alone has no protection from stray dogs or predators. He could also become tangled and hurt or scare himself.

RULE 5: KEEP A FRIEND

Goats are herd animals. A solitary goat becomes lonely without constant companionship. Usually, this means the goat owner should keep at least two goats. However, a goat will bond with almost any species of stall mate, including a horse, sheep, or cow.

Left completely alone, a goat will develop a number of negative behaviors. Some goats cry. Lonely Nubians in particular have very loud voices. A crying goat quickly becomes as annoying to your neighbors (and you) as a barking dog.

A PET FOR YOUR HORSE

Goats make good companions for horses. Just as a goat gets lonely, a single horse pines for friendship. The phrase "to get his goat" originated from the practice of stealing the goat companion of a racehorse before the

Left to their own devices, goats run wild just for fun—or tussle to decide who is higher in herd status. *Jen Brown*

Goats and horses have a natural relationship that dates back to the early days of horseracing, when small, cheap, and easy-to-transport goats were kept as stall companions to calm down high-strung racehorses.

With pet goats, it's best to have at least two for the sake of the goats. A solitary goat is often an unhappy, noisy, troublesome goat. *Shutterstock*

Wethers as Pets

The most common pet goat is a castrated male known as a wether. This popularity is mostly by process of elimination. A breeding buck smells bad and has all sorts of unappealing behaviors (such as peeing on his face!). A doe needs to be milked and comes into heat every fall. Both tend to be prohibitively expensive.

The best place to buy a wether is from a reputable breeder. A breeder may charge as much as $150 but this typically gets you a neutered, disbudded, vaccinated animal plus after-sale support. You can also buy wethers from auctions or private parties, but buyer beware. Free or cheap goats can be either a tremendous bargain or a heartbreaking mistake.

The wether has special diet considerations. Unlike a buck or a doe, a wether does not expend energy on mating or milking. An overfed wether quickly becomes fat. Limit his grain and treats. A wether is also prone to urinary calculi, a potentially fatal condition. To help prevent, supplement his feed with ammonium chloride—and keep his weight down!

Goats are playful and need to be active to stay healthy. An outside pasture or yard can be made more appealing by providing distractions. *Jen Brown*

costume classes at shows are fun for goat exhibitors and spectators alike. Elaborate or simple playgrounds provide exercise for the goat and keep the owners amused.

AGILITY OBSTACLES

- **Balance beam:** A plank suspended a few inches or a foot off the ground that the goat must walk across without stepping off.
- **Bridge:** A small bridge that the goat must cross.
- **Jump:** A raised bar, crossed logs, or other material that the goat must leap over.
- **Slalom:** A zigzag course made of poles that the goat must weave through without skipping or knocking any over.
- **Teeter-totter:** A plank balanced on a central pivot so that the plank tips as the goat crosses.
- **Tent:** A tent the goat and owner must enter.
- **Tire hanging:** A tire suspended in the air that the goat must jump through.
- **Tires:** A course made of tractor or truck tires, placed flat on the ground, that the goat must step into and out of as it crosses.
- **Tunnel:** A vinyl or hoop tunnel that the goat must pass through.
- **Water hazard:** A kid's wading pool, stream, or large puddle the goat must step into.

race. Any size goat can be paired with a horse. The advantage to using a full-size breed is that the goat is closer in size to the horse.

Introduce the goat and horse gradually by penning them in adjoining stalls. Use caution during the first few weeks, while the animals are adjusting to each other. Be aware of possible problems, and be ready to separate the animals if necessary. Horns on a goat can be dangerous for the horse. Some goats like to chew on the horse's tail. An aggressive horse may kick the goat or pick it up and throw it.

Fencing for horses is frequently not goat-proof. Adding lower electric strands or making a smaller fenced area for the goat may be necessary.

FUN STUFF

One of the joys of owning goats is their playful manner. Goats love attention, so the tricks and games you can play are limited only by your imagination. Some of these games have become formal competitions at fairs. Agility and

PACK AND HARNESS GOATS

Goats can pull a cart or carry a pack. Itinerant traveler Ches McCartney, otherwise known as the Goat Man, famously wandered the United States with a wagon pulled by a team of goats from the 1930s to the 1960s. Some travel companies use pack goats for expeditions. Goats are perfect for moving about rough land that other animals find impossible to maneuver.

Size is the main factor limiting the pulling or carrying capacity of a goat. A well-conditioned pack goat can carry up to 25 percent of its weight. The same animal should not be asked to pull a cart more than one-and-a-half times its weight.

Training a goat to work is neither easy nor quick. The effort pays off with time and patience. Goats love attention and company, so those trained to pull or carry will soon enjoy the activity as much as you do.

Empty buckets sometimes get used as toys, but goats don't like being in the dark and will become spooked if the bucket gets stuck. *Jen Brown*

A pack goat can pull a cart for exhibition or work around the farm. Hoppy the cart goat was always popular at the Washington County Fair in Minnesota. *Carol Amundson, Happy Land*

149

CHAPTER 16

• • • • • • • • • • • • • • • • • •

SHOWING
YOUR
GOATS

The heart is like a goat that has to be tied up.
—South African proverb

A child interacts with goats on display. *Shutterstock*

I once believed that showing animals was all about vanity. Admittedly, when you have a particularly nice animal that seems unbeatable, it is hard to remain humble. But the best reason to show has nothing to do with pretty ribbons, trophies, or champion legs. It's to compare your animal with other goats. Getting other people's opinions about the goats in your barn and sharing war stories with other breeders is an education in and of itself.

My first two Nubians, Celeste and Menolly, were meant to be milk and hobby-farm animals. I had no interest in showing goats. Linda Libra convinced me to bring the "girls" up for a day at the East Central Dairy Goat Association Show. I labored to clip just enough hair (but not too much!) so that the kids looked neat.

The weather was cold and rainy. Goats and people seemed to be everywhere. Talk was indecipherable. My kids looked shaggy compared to the other, perfectly trimmed animals. Several breeders kindly commented on how smart I was to leave them "fuzzy" so they wouldn't

catch a chill. The whole day was confusing but fun. When Celeste took first-place senior kid, I was hooked on the world of goat shows.

The world of showing goats offers many possibilities. Shows run the gamut from "fuzzy trailer shows"—where the goats are unclipped, brushed, and shown right out of the owner's trailer—to public events where the goat and handler are at their cleanest and best. County and state fairs hold shows that pay modest premiums for animals shown in either open class or some of the youth-centered programs such as 4-H and FFA. Breed associations sponsor shows. Most of these organizations hold at least one national show each year. In the case of pet goats, sometimes fairs have fun classes in which the goat and owner show together in a costume contest, obstacle course, or cart exhibition.

FINDING SHOWS

Contact the breed registry for whichever type of goat you want to show. Check the fair's premium books or websites. Become a member of a local goat club to get news of upcoming shows and events.

Once you are integrated into the caprine network, finding shows is fairly simple. Unfortunately, in some

150

areas, there are no nearby shows. Then you need to decide if you want to travel. For more and more states, there is an added cost to crossing state lines with goats. State regulations may require travel and health documents, which are discussed in chapter 3.

TO CLIP OR NOT TO CLIP?

Fitting requirements vary depending upon the show, the type of goat being shown, and even the weather. In some areas, the goat is lightly trimmed; some find a ⅜-inch clip preferable; and sometimes a meat animal is "slick sheared" or clipped like a dairy goat. Some breeds of goats should not be clipped before the show. In the case of fleece shows, the goat may not go at all, only her wool. Miniature goats and meat goats are often shown unclipped.

Each exhibitor brings feeders and other gear to make the goats feel comfortable in their temporary home. *Carol Amundson*

Goat Show Packing Checklist

Livestock and Gear	Show Supplies	Personal Gear
Goats	Health papers	Tent
Feed	Registration papers	Cot
Hay	Show collars	Sleeping bag
Hay and grain feeders	Show whites	Pillow
Fasteners for feeders and display	Display items	Toiletries
Bedding straw or shavings	Farm sign	Food/snacks
Rake	Business cards	Coffee pot and supplies
Broom		Street clothes
Milking equipment		Black shoes
Milk stand		Buckets
Pasteurizer		First-aid kit (human)
Clippers		
Hoof shears or knife		
Grooming supplies		
Shampoo and conditioner		
Tools		
First-aid kit (caprine)		

Agility obstacles are another fun category at goat shows. Slalom poles and a teeter-totter are but a few of the challenges placed before the goats and their handlers. They are judged by time and skill. *Carol Amundson*

Some local fairs have a costume show, a fun category judged on originality, comfort, and ease of movement for the animals. Normally the goat and exhibitor must both be in costume. *Carol Amundson*

JUDGES, JUDGING, AND KEEPING IT ALL IN PERSPECTIVE

Shows are subjective. Each registry tries to make showing more objective through their scorecards and judges' training, but it isn't easy to select the best goat from a lineup of beautiful animals. The doe that stood first yesterday may not place as high today and vice versa. One judge's opinion is just that, an opinion, no matter how many years he has been judging or what scorecard he is using. Before you start to show your goats, I strongly recommend going to shows in your area. Talk to exhibitors. Watch the judges and listen to their reasons.

Shows are fun and exciting. They can also be exhausting, frustrating, and downright discouraging. Each show is a new experience. One of my friends, who has been showing goats since her son was in 4-H (he now judges and has his own children), has one of the best show perspectives I have ever encountered. Her philosophy is "We show what we've got." She continually works toward breeding the best goats she can while raising animals she likes to have around.

Nubians wait their turn in the show ring. *Shutterstock*

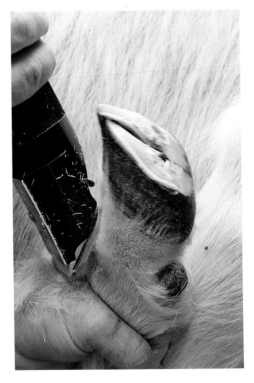

Above and right: Clipping for the show is usually done beforehand at home, but having the clipper and lube available at the fair helps for last minute trimming of pasterns or udders. *Jen Brown*

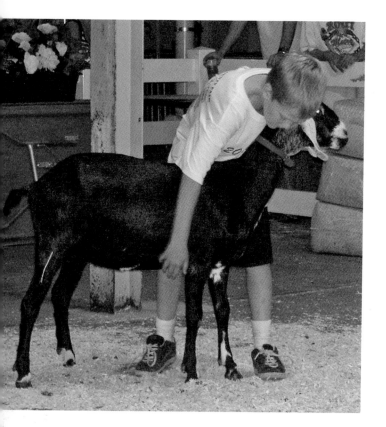

Watching shows helps learn show techniques and etiquette. Goats are moved at the judge's direction. As the goats are walked around the ring or stopped for inspection by the judge, exhibitors should always be aware of how the goat appears to the judge. *Carol Amundson*

Anna Thompson, of Legendairy Goats, milks out her Toggenburg doe following competition at the Minnesota State Fair. Milking at the fair makes for great public relations. *Carol Amundson*

Showmanship Tips

- Train your goat to lead and stand properly before the show.

- Walk your goat at a steady pace. Don't run or move too quickly.

- Watch the judge. Always know where the judge is and what is expected of you.

- When possible, walk your goat into the showing area before the show so it can become familiar with the surroundings.

- Hold your goat's head high with the collar held under the chin. A choke-chain collar is most commonly used in the show ring. This is not a safe collar to leave on an animal, so be certain to remove it after the show.

- Keep the goat between you and the judge. This is known as the "peanut butter sandwich." The goat is the peanut butter, and you and the judge are the bread.

- Move in front of the goat (never behind) when turning.

- Be aware of your goat and set it square when standing still. Don't over-handle your animal. Once it is set, avoid distracting the judge from looking at your goat with a lot of movement.

- Don't crowd the goat ahead of you. Keep a few feet between your animal and the next one so the judge can look at each goat properly. This buffer also prevents tail biting and other negative contact between unfamiliar goats.

- Know your goat. The judge may ask for the animal's age or birth date, kidding date, or number of kiddings. For showmanship class, be prepared to answer more technical questions about management, parts of the goat, or how you feed, vaccinate, and worm your project animal.

MARKETING YOUR GOATS AND THEIR PRODUCTS

Goats and their products can be marketed through many venues. *Shutterstock*

They who don't keep goats and yet sell kids, where do they get them?

—Spanish proverb

Live animals, milk, meat, hair, and hide can all be marketed, depending on your supply and local regulations. Here is an overview of some sales venues and how to approach them.

AUCTION OR SALES BARN

The sales barn or local livestock auction is a way to sell animals with minimal fuss. You simply drop the animals off and receive payment at the end of the auction. Live auctions sell goats by the pound or by the head. In the case of stockyards with buyers on the premises, you may even get their market price paid to you directly at the time of delivery, without having to wait to sell the animal first. Rarely do you receive top price this way, but it is a good way to dispose of culled animals.

Local newspapers and feed mill bulletin boards are good resources for finding out when auctions will be held. Some sales are weekly occurrences, others monthly. The auctioneers take a commission from your sale and may also charge a stall fee. Specialty auctions, on the other hand, are set up specifically to attract breeders with papered goats or other livestock.

DIRECT MARKETING

Direct marketing to individuals is probably the most common method of selling live goats. Goats and their products are specialty items with a market best reached through targeted advertising rather than mass marketing.

SIGNS AND BULLETIN BOARD NOTICES

An attractive farm sign identifies your location and attracts drive-by business. Most agriculturally zoned properties allow sign advertising on their property without community approval.

Bulletin boards at the feed mill, grocery store, or pet store are other places to post sales bills. If you have a product to market beyond live animals, you are limited only by your imagination. The grocery store, the health food market, and ethnic gathering spots are all worth trying.

PRINT ADVERTISING

Print ads are paid by the word, line, or page layout (such as a quarter of a page). Be sure to track which ads bring in buyers before spending too many advertising dollars in this area. At minimum, your ad should contain the basics of what you are selling (goats, yarn, meat, etc.) and contact information. I include my farm name, phone number, and address or general location. If you have a website, this is a great opportunity to direct people to the site.

Local newspapers will get the ad to buyers right in your area. Most papers have an online presence ad and will list your ad in both places. Specialty farm newspapers target your market even more precisely.

Magazines can be useful for a general monthly farm ad. Don't limit yourself to just the industry magazines. Goats are popular with back-to-the-landers, homesteaders, and anyone interested in organic and sustainable products.

Local goat associations often have a club newsletter. Here in Minnesota, I enjoy the *Gopher Goat Gossip* to find out what is happening on the state dairy goat scene. It is also a source of inexpensive advertising.

INTERNET MARKETING

The Internet has made marketing a whole new world. It can be as inexpensive or as costly as you want, depending on your approach and Internet skills. The ability to link directly to your target market is a big advantage.

A well-designed personal website advertising your farm allows you to reach buyers with as much or as little information as you have the time and inclination to provide. You can share your farm's history, photos of your farm and animals, pedigrees, your breeding philosophy, and more.

Cybergoat, the Goat Connection, GoatWeb, and GoatFinder are just a few of the goat-specific websites that have places to list your farm or advertise products and animals.

Some free farm bulletin boards exist online. Others charge fees, the same as print media. This type of marketing can work well if you place frequent ads and individualize the animals you have for sale.

Craigslist is an online classified ad service that offers no-cost advertising. It's a great place to sell caprine

A pen containing pet goats is always a hit with visitors to our market garden. *Carol Amundson*

An enterprising youth group at the fair tells people about goats and lets them try their hand at milking. *Carol Amundson*

products such as goat-milk soap, hide, hair, and even leftover equipment. While pet sales are prohibited by Craigslist rules, it is legal to list livestock for sale.

FARM STAND

A farm stand—at a farmer's market, at a craft show, or in your yard—reaches people looking to buy direct from the producer. You can sell raw fiber and hides or processed items such as goat-milk soap and cashmere scarves. It is not common to sell livestock at a stand, and local laws regulate the sale of products for human consumption, including goat milk and meat.

ASSOCIATIONS AND AGRICULTURE COMMUNITY LINKS

Your state Department of Agriculture may have marketing programs that provide low-cost or free marketing for producers. Contact your local County Extension Office, listed in the telephone book under Agricultural Services in the county listings, or your state Department of Agriculture to investigate whether your state offers sale opportunities.

SHOWS

Have a presence at livestock shows, and keep track of your goats' wins and placings. An attractive display at these events is a useful marketing tool. This method is time-

Dairy products *(top, left)*, goat meat *(top, right)*, and fine fiber *(above)* are some of the products a farmer can obtain from goats. *Shutterstock*

consuming and costly. It also reaps rewards for the diligent breeder who wishes to sell breeding and show stock. Goat shows allow breeders to talk to one another, see animals at their best, and hear the opinions of trained judges.

SALES LIST

Compile a list of the goats you have for sale, including each animal's age, sex, price, and other important details. Distribute your sales list at fairs and shows, and be sure to collect the sales lists of other breeders. The breeder with whom you exchange lists may very well be a customer for outcross animals at a later date. If you don't have any goats for sale during a particular show, collect names and addresses so you can forward information as animals become available. Remember, the goat community is relatively small, so keep cordial relations with as many breeders as possible. Word-of-mouth exchanges can be very helpful—or they can ruin your reputation. I regularly get referrals from other goat breeders who either have no animals available or don't raise the same breeds I do. In the same way, I pass buyers on to others.

KID PICKUP OR MARKETING POOL

Farmers sometimes pool their goat kids and other meat goats for holiday markets. Called a kid pickup or marketing pool, this event occurs in the spring for Easter and Passover in the upper Midwest, but timing may vary in different parts of the country. In my area, Daniel Considine of Sunshine Farms has contacts within the goat community. Each spring, Dan sends out letters letting farms know when he will be collecting animals. Growers meet at a predetermined location with their excess bucks, wethers, and other kids suitable for meat.

Producers can also get together a group of similar goats to sell to a volume buyer. The animals are weighed at a central pickup point. Some co-ops use this approach and negotiate prices based on the higher volume of goats.

A kid pick-up is a valuable community effort to sell excess dairy or fiber kids for meat. *Shutterstock*

COOPERATIVES

Goat cooperatives have been formed for selling caprine items such as milk, cheese, meat, and fiber. If you raise a small herd of goats, it can be tough to keep up with market demand. Goat breeding is seasonal in nature, and an individual farm creates a relatively small amount of saleable product. Forming a cooperative helps to increase product supply while evening out some of the seasonal variation. It also allows co-op members to pool resources for advertising, processing, and shipping.

SALES ETHICS

"Decisions are easy when your ethics are clear."
—Walt Disney

Marketing your goats can be a hobby or a business—or just a byproduct of having more goats than you need. Whatever the case, treating your animals, your products, and your buyers as you would want to be treated isn't just ethical, it's good business. Ethical sales spur repeat business.

If you belong to a goat registry, it most likely has a code of ethics or a trade policy for selling goats. These guidelines can be useful for creating your own set of rules on your farm. Outlining your promise to your customer in a simple sales agreement can go a long way toward avoiding misunderstandings:

Disclose any problems up front. No animal is perfect. If there are defects, be straight with potential buyers. If a doe has been a disappointing milker, a poor mother, or problematic in some other way, provide the information up front. You may still make the sale. I have bought problem goats because they had other features that appealed to me.

Educate new owners. When confronted with a brand-new goat owner, expect to answer lots of questions. Tell them the basics of feeding, housing, and caring for their new livestock. This can save you time answering panicked calls later and may save the goat you are selling from injury—or death. Willingness to provide follow-up advice is the sign of a good breeder. If you just want to unload a low-quality goat, take it to the sales barn.

Above and opposite: Take care to follow a code of ethics when selling a goat to a buyer. *Shutterstock*

Focus on a "fit" and talk enough with your buyers to know what they want. You want to sell your goat or product. However, it doesn't make sense to sell an animal to anyone whose needs won't be met by the sale. Spur-of-the-moment sales are never a good idea. In the best-case scenario, the animal returns home none the worse for wear. At other times, goats or their products get a bad name, quality animals end up in the sales barn, and you get negative word-of-mouth.

Identify the animal clearly by making sure tattoos, ear tags, or other IDs are present and legible. Premises ID and regulations aside, it is helpful to know which goat you are selling. Tattoos can fade—or possibly were not there to begin with. Ear tags pull loose. The goat leaving your farm should be the one you want to sell and the one the buyer wants to own. Accidents happen on the best of farms. A simple check at the time of loading prevents problems down the line.

Maintain good records and provide them to buyers when necessary. Pet-goat owners may want nothing more than the age of their goat. Buyers of breeding or show stock require more—everything from registration papers to show and production records. Have this information in an organized and easily accessed system.

This practice is especially important when selling breeding services. Provide the service memorandum at time of service. Disclose any problems with papers before the sale or breeding is finalized.

Never repeat negative gossip about another farm— and think twice about telling a negative anecdote about a firsthand experience. It is good practice to worry only about your own reputation, not that of your competition. Negative information overheard about another farm may be true—or it may be the result of a misunderstanding. Repeating stories rather than focusing on the strengths of your operation can reflect as poorly on you as on the one whose faults you expose.

Sell healthy animals. To some breeders, assuring that goats are healthy means testing for known diseases, raising kids using CAE prevention techniques, and maintaining a closed herd. To others, it simply means making certain that the goat exhibits no obvious signs of illness and is current on vaccines and hoof trimming. Spell out your health management practices to prospective buyers. Some sellers give vaccines, worming, and hoof trimming at the time of sale to show the buyer how those tasks are done. This is also tangible proof that you have done them.

Use responsible management and proper care, including housing, feed, health care, and maintenance. Your goats and the products provided by them are only as good as the care they receive. Keeping up-to-date on goat management is an ongoing process. Regardless of whether your management style is conventional, organic, or somewhere in between, it is good business to take good care of your herd.

. . . .

MANAGEMENT CALENDAR

This calendar recommends general management activities for goats. Owners should modify the calendar to fit their breeds, management style, and environment.

FALL	Bucks	• Clip excess belly hair. • Examine penis and testicles for injuries or inflammation. • Shorten or remove scurs before breeding. • Treat as recommended by semen processor if the buck's semen will be collected. • Turn out bucks with does at a ratio of 1:25 to 1:50. • Use a raddle harness on bucks before pasture breeding for improved tracking of breeding dates.
	Does	• Continue to flush does for two to three weeks after they are penned with a buck. • Culture milk to pick up subclinical mastitis. • Examine udders for signs of mastitis or injury. • Plan dry-treatment if mastitis has been a persistent problem; treat dry does when kids are weaned or milking stops. • Reduce grain for does that are too fat. It is easier and safer to change body condition before dry-off.
	Kids	• Choose replacement does and bucks. • Evaluate final kid crop. • Wean kids.
	Herd	• Treat any lice or skin disorders. • Cull unsound or inferior animals. • Determine body condition scores. • Increase grain for two to three weeks before and after breeding. • Give supplemental selenium and/or copper, if necessary, two to four weeks before breeding. • Record breeding dates. • Monitor internal parasites by testing stool samples. • Perform any dehorning procedures after frost if season permits. • Plan winter supplemental feeding program. • Test herd for diseases of concern. • Cull or isolate test-positive animals from the clean herd. • Treat for intestinal parasites as necessary after a hard freeze or before moving to new housing.

Shutterstock

Fall	Herd	• Trim feet and monitor for foot rot before rainy season. • Vaccinate for tetanus and enterotoxemia (CD/T) unless your veterinarian recommends a different schedule.
	Housing	• Clean and disinfect pens and barn before the start of bad weather. • Evaluate housing and repair as necessary, paying special attention to leaky roofs and drafts.
	Pastures	• Evaluate range and forage conditions. • Move herd to fresh green pasture, if available, before pasture breeding begins. • Soil-test pasture and supplement as needed.

WINTER	Bucks	• Remove bucks from breeding pen or pasture after does have settled. • Supplement diet to regain body condition lost during rut.
	Does	• Delouse fiber goats at shearing if necessary. • Dry off and dry-treat udders sixty days before due date. • Give vaccine boosters three to five weeks prior to due date. • Gradually increase quality of hay and amount of grain fed to late-pregnancy does. • Perform pregnancy testing; the best window for ultrasound evaluation is forty-five to sixty days post breeding. • Shear or comb fiber goats three to six weeks prior to kidding. • Sort pregnant does from open does. • Supplement with selenium, if necessary, prior to due date. • Pen unbred does with a buck for clean-up breeding.
	Kids	• Breed late kids after seven months of age or at 75 pounds body weight. • Castrate any late bucklings to be raised for meat. • Tattoo or tag late kids. • Vaccinate late kids according to schedule.
	Herd	• Check for lice and delouse if found. • Monitor body condition of all goats and adjust supplemental feeding as required. • Prepare sales materials and distribute sales lists. • Cull unsound or inferior animals. • Stock kidding supplies. • Trim feet as needed. • Clean equipment such as clippers so they are ready for kidding and show seasons. • Watch older, younger, and other vulnerable animals for chilling.
	Pastures	• Ensure kidding pastures have adequate shelter for freshening does. • Evaluate forage and pasture conditions. • Select kidding pasture if range kidding.

Jen Brown

SPRING	Bucks	• Assess bucks that are available for breeding. • Begin looking for replacement bucks. • Give vitamin E/selenium (BoSe) in selenium-deficient areas.

	Does	• Clip udders on non-fiber goats at least one week prior to due date.
		• Freeze heat-treated colostrum for emergency use and CAE prevention.
		• Perform CAE prevention as necessary:
		   o Tape doe's teats one week before due date.
		   o Segregate known CAE-positive does from the herd.
		   o Remove kids from doe immediately after birth.
		   o Feed heat-treated colostrum, pasteurized milk, CAE-free milk, or milk replacer.
		• Shear or comb fiber goats three to six weeks prior to kidding.
		• Supplement lactating does to maintain milk production, increasing feed gradually.
		• Supplement non-dairy goats for at least four weeks post-kidding.
		• Worm does at freshening or one to two weeks following kidding.

	Kids	• Check the doe's teats and start milk flow so the kid may nurse easily. • Feed colostrum within the first couple of hours after birth. • Identify kids. • Keep records of each freshening. • Neuter males that are not being raised for breeding. • Disbud kids. • Monitor kids raised on the dam to be sure each is getting enough to eat. • Pen kids with individual does for dam-raising when possible. • Send registrations for new animals. Some registries, such as ADGA, have lower fees for early registration. • Start coccidia prevention or test feces by three weeks of age. • Supplement weak or slow-growing dam-raised kids with bottle feeding.
	Herd	• Monitor internal parasites by testing stool samples. • Treat for worms before cleaning the barn or moving the goats to new pasture when testing shows high levels of worms. • Shear non-pregnant fiber animals.
	Housing & Equipment	• Clean and bed kidding areas several days before the doe's due date (if not completed in fall). • Renew trailer licensing and affix tabs. • Clean trailer or hauling equipment for market and show season. • Order fly-control products. • Sort show supplies and tack to be sure everything is in order.
	Pastures	• Check and repair fencing. • Check outside pens before moving animals. Repair fence as required.

Shutterstock

SUMMER	Bucks	• Assess bucks that are available for breeding. • Continue looking for replacement bucks. • Market excess bucks and cull those that don't sell. • Give vitamin-E/selenium (BoSe) in selenium-deficient areas in late summer.
	Does	• Cull does that have reproductive problems or ill health. • Discontinue supplemental feeding to does that are at least four weeks postpartum and aren't being milked for dairy.
	Kids	• Evaluate kid crop to decide which will be sold and which kept for replacement. • Monitor internal parasites by testing stool samples. • Continue coccidia prevention or start testing feces by three weeks of age. • Vaccinate at four, eight, and twelve weeks of age or as recommended by veterinarian. • Worm at six to eight weeks or as necessary. • Tattoo or tag kids at four to six weeks of age.

	Herd	• Clip or bathe goats to remove dead skin.
		• Check for external parasites and ringworm.
		• Keep animals hydrated and cool with water and ventilation during hot-weather hauling.
		• Maintain adequate shade for animals during hot weather.
		• Show season:
		   o Do not overfill the udder on does, which can cause mastitis.
		   o Make feed changes gradually.
		   o Double-check show supplies.
		   o Send show entries on time.
		• Test stool samples for internal parasites.
		• Treat for worms if necessary.
		• Trim feet as necessary.
		• Watch for foot rot in wet conditions.
		• Make sure clean water is available at all times. Water usage increases due to lactation and hot weather.

	Housing	• Continue fly-control measures. • Keep pens well-bedded or clean frequently to prevent illness.
	Pastures	• Rotate pastures every several weeks, if possible. • Repair fencing and pens before moving animals. • Watch pastures for poisonous plants and remove if present.

Shutterstock

ASSOCIATIONS, CLUBS, AND REGISTRIES

Bring me a bowl of coffee
before I turn into a goat.
—Johann Sebastian Bach

Groups for people interested in goats are forming and disbanding all the time. This list is by no means comprehensive, but at the time of publication, I personally verified the current information of each organization. Most clubs use the phone and address of their secretary as a contact. Since the secretary is usually elected, the address of clubs may change more frequently than the address of organizations such as registries, which have a permanent address.

Alaska Mini Goat Cache
4228 Country Fair Dr
Wasilla, AK 99654
(907) 376-4951
www.alaskaminigoatcache.tripod.com
alaskaminigoatcache@yahoo.com

Alberta Mohair Producers Association
www.albertamohairproducers.ca

Alpines International
Tina Antes, Secretary
7195 County Rd 315
Silt, CO 81652
www.alpinesinternationalclub.com
goat@sfalpines.com

American Angora Goat Breeders Association
PO Box 195
Rocksprings, TX 78880
(830) 683-4483
www.aagba.org

American Boer Goat Association
1207 S Bryant Blvd, Suite C
San Angelo, TX 76903
(325) 486-2242
www.abga.org
info@abga.org

American Cheese Society
455 S Fourth St, Suite 650
Louisville, KY 40202
(502) 583-3783
www.cheesesociety.org
acs@hqtrs.com

American Dairy Goat Association
PO Box 865
Spindale, NC 28160
(828) 286-3801
www.adga.org
info@adga.org

American Goat Society
735 Oakridge Ln
Pipe Creek, TX 78063
(830) 535-4247
www.americangoatsociety.com
agsoffice@earthlink.net

American Golden Guernsey Association
www.guernseygoats.org

American Kiko Goat Association
Jean Thomure, Secretary
295 Cardinal Ln
Trenton, GA 30752
(706) 657-8649
www.kikogoats.com
2meracre@tvn.net

American LaMancha Club
Deb Macke, Secretary
W1445 Old 14 Rd
Hawkins, WI 54530
www.lamanchas.com
alc@lamanchas.com

American Livestock Breeds Conservancy
PO Box 477
Pittsboro, NC 27312
(919) 542-5704
www.albc-usa.org

American Meat Goat Association
PO Box 676
Sonora, TX 76950
(325) 387-6100
www.meatgoats.com

American Nigerian Dwarf Organization
1510 Bird Rd
Independence, KY 41051
www.andda.org

Appalachian Goat Association
http://members.aol.com/agagoat/club.html
AGAgoat@aol.com

Arapawa Goat Breeders
Al Caldwell, Registrar
88 Gorham St
Rehoboth, MA 02769
(508) 252-9469
www.arapawagoat.org
alspond@comcast.net

Arizona State Dairy Goat Association
Denise Thomas
13805 E Appleby Rd
Chandler, AZ 85249
www.asdga.com

Arkansas Boer Goat Association
Jerome Gentry Jr
152 Mitchell Lane
Beebe, AR 72012
(501) 882-0181
www.arkansasboergoatassociation.com

Arkansas Goat Producers Association
Debbie Taylor
498 Bluff Top Rd
Dover, AR 72837
www.freewebs.com/arkansasgoatproducers
arkansasgoatproducers@gmail.com

Big Country Goat Producers
PO Box 5211
Abilene, TX 79608
www.bigcountrygoatproducers.com

Canadian Cashmere Producers Association
www.canadiancashmere.ca
info@canadiancashmere.ca

Canadian Goat Society
2417 Holly Lane
Ottawa, ON K1V OM7
(613) 731-9894
www.goats.ca
cangoatsoc@travel-net.com

Canadian Livestock Records Corporation
2417 Holly Lane
Ottawa, ON K1V OM7
(613) 731-7110
www.clrc.ca
clrc@clrc.on.ca

Capricorn Cashmere
www.capcas.com
Info@CapCas.com

Cascade Boer Goat Association
www.cascadebga.org
info@cascadebga.org

Cascade Pygmy Goat Association
PO Box 1282
Graham, WA 98338
www.cpga-pygmy.com

Colorado Dairy Goat Association
Marilou Webb
459 Old St. Vrain Rd
Lyons, CO 80540
www.colodga.org
southforklm@aol.com

Colorado Pygmy Goat Club
Sandra Hoihjelle
12925 Holmes Rd
Colorado Springs, CO 80908
(719) 495-5760
www.copygmygoatclub.com

Colored Angora Goat Breeders Association
Shelley Aikens, Treasurer
14663 Evans Creek Rd E
Rogue River, OR 97537
(541) 582-3705
www.cagba.org
ddnorton1@msn.com

Delta Dairy Goat Association
Mary McMahan, Secretary
44465 Marie Dr
Hammond, LA 70403
www.angelfire.com/la3/DDGA

East Texas Goat Raisers Association
PO Box 1128
Big Sandy, TX 75755
(903) 721-3347
www.etgra.com

Eastern Cashmere Association
Chuck Vailes
Silver Branch Farm
1506 Sangers Lane
Staunton, VA 22401
www.easterncashmereassociation.org
chuck@vailes.com

Empire State Meat Goat Producers Association
PO Box 924
Corning, NY 14830
(315) 662-3250
www.esmgpa.org
mysticvalleyfarms@frontiernet.net

Evergreen Pygmy Goat Association
Staci McStotts
23727 – 19th Ave NE
Arlington, WA 98223
www.geocities.com/evergreenpygmygoatassociation
criticschoice@wavecable.com

Florida Dairy Goat Association
Jan Noble, Secretary
2450 212 Ct NE
Williston, FL 32696
(352) 528-2022
www.fdga.org
conip@hughes.net

Florida Meat Goat Association
Diane Strickland, Treasurer
1734 County Rd 227A
Oxford, FL 34484
www.fmga.org

Four States Boer Goat Association
Jennifer Hawthorn, Secretary
Arkadelphia, AR 71923
(870) 246-6353
www.4sbga.com
jen@cedargrovefarms.com

Georgia Dairy Goat Breeders Association
John Latimer, Secretary
1540 McRee's Mill Rd
Watkinsville, GA 30677
www.georgiagoat.com
LLacres@juno.com

Georgia Meat Goat Association
1131 Treadwell Bridge Rd
Statham, GA 30666
www.members.tripod.com/gmga
vskinner@bellsouth.net

Goat Lovers of Western Massachusetts
61 South Washington St
Belchertown, MA 01007
(413) 283-4815
www.glow-goatclub.50megs.com
glow_glu@yahoo.com

Gulf Coast Boer Meat Goat Association
PO Box 58
Santa Fe, TX 77510
www.angelfire.com/tx6/gcbmga

Heartland Nigerian Dwarf Goat Association
Mary Segal, Secretary
www.hndga.org
msegal@itlnet.net

Idaho Springs Dairy Goat Association
Kristin Forcier, President
www.isdga.com
4forciers@gmail.com

Illinois Meat Goat Producers
www.ilmeatgoat.org
vicky@deerhavenfarms.com

Indiana Dairy Goat Association
Bonnie Robbins, Secretary
9076E 300S
Greensburg, IN 47240
(812) 934-3010
www.idga.net
borobbins@seidata.com

International Boer Goat Association
PO Box 310
Bonham, TX 75418
(903) 640-4242
(877) 402-4242
intlboer@intlboergoat.org

International Dairy Goat Registry
Route 1 Box 173
Milo, MO 64767
(417) 884-2455
www.goat-idgr.com
idgregistry@gmail.com

International Fainting Goat Association
2455 Deanburg Rd
Pinson, TN 38366
www.faintinggoat.com

International Goat Association
1 World Ave
Little Rock, AR 72202
(501) 454-1641
www.iga-goatworld.org
goats@heifer.org

International Kiko Goat Association
PO Box 677
Jonesborough, TN 37659
(888) 538-4279
www.theikga.org

International Nubian Breeders Association
Caroline Lawson, Secretary
5124 FM 1940
Franklin TX 77856
(979) 828-4158
www.i-n-b-a.org
secretary@i-n-b-a.org

International Sable Breeders Association
3794 Deckerville Rd W
Elma, WA 98541
www.sabledairygoats.com
webmaster@sabledairygoats.com

Iowa Meat Goat Association
Daniel Palmersheim, Secretary
2241 220th St
Manchester, IA 52057
www.iowameatgoat.com
info@iowameatgoat.com

Junior Meat Goat Show Circuit
1004 Virginia Circle
Big Lake, TX 76932
(325) 884-3645
www.jmgsc.com
jmgsc@jmgsc.com

Kansas Meat Goat Association
Vanessia Ochs, Secretary
1550 Rock Creek Rd
Ottawa, KS 66067
(785) 418-6530
www.kmga.net

Kentucky Goat Producers Association
Tom Garrity, Secretary
13207 Rehl Rd
Louisville, KY 40299
(502) 266-8531
thomasgarrity@bellsouth.net
www.kentuckygpa.com
kygoat@fewpb.net

Keystone Pygmy Goat Club
Maggie Leman
1205 Olive Branch Rd
Durham, NC 27703
www.keystonepygmygoatclub.com
maggidans@msn.com

Kinder Goat Breeders Association
PO Box 1575
Snohomish, WA 98291
www.kindergoats.com

Manitoba Goat Association
Susan Frazer, Treasurer
General Delivery
Beulah, MB R0M 0B0
www.manitobagoats.com
president@manitobagoats.ca

Maryland Dairy Goat Association
Marci Waterman
15609 Old Frederick Rd
Emmitsburg, MD 21727
www.marylanddairygoat.org
secretary@marylanddairygoat.org

Maryland, Pennsylvania, & West Virginia Meat Goat Producers Association
Pam Adams, Treasurer
400 Bridgestone Dr
Eldersburg, MD 21784
(443) 802-3734
www.meatgoat.biz
pamela.j.adams@eds.com

Meat Goat Producers Association
Laura Sandness, Secretary
(620) 879-2050
www.mgpa-sek-neo.com
sscowdogs@yahoo.com

Michigan Boer Goat Association
Manuel Ibarra
4577 Gresham Hwy E
Potterville, MI 48876
www.michiganboergoat.org
inquiry@michiganboergoat.org

Michigan Dairy Goat Society
8653 Chippewa N
Coleman, MI 48618
(989) 465-1506
www.mdgs.org
webnanny@mdgs.org

Mid-South Goat Masters
Clarence Pemberton
202 Nichols Dr
West Monroe, LA 71291
www.mid-southgoatmasters.com
info@mid-southgoatmasters.com

Midwest Goat Producers
Vicky Wetzel, Deer Haven Farms
13505 Dix Texico Rd E
Texico, IL 62889
www.midwestgoatproducers.org

Miniature Dairy Goat Association
PO Box 7244
Kennewick, WA 99336
(509) 591-4256
www.miniaturedairygoats.com

The Miniature Goat Club
8370 Abbey Lane W
Wilhoit, AZ 86332
(928) 776-0092
www.theminiaturegoatclub.com
info@theminiaturegoatclub.com

The Miniature Goat Registry Inc
7870 Shelburne Rd W
Wilhoit, AZ 86332
(928) 442-9127
www.tmgronline.org
reg@tmgronline.org

Miniature Silky Fainting Goat Association & Registry
22105 Countryside Lane
Lignum, VA 22726
(540) 423-9193
www.msfgaregistry.com

Minnesota Dairy Goat Association
Krista Matson, Membership Chair
12104 State Hwy 70
Pine City, MN 55063
(320) 629-0081
www.minnesotagoats.org

Mississippi Dairy Goat Association
Chris Holland
PO Box 1162
Belmont, MS 38827
www.missdairygoat.com

Mississippi Goat Association
www.mississippigoatassociation.org
info@mississippigoatassociation.org

Missouri Junior Boer Goat Association
www.mjbga.org

Montana Mohair Producers Inc
Valerie Hogan
73 Poverty Flats Rd
Eureka, MT 59917
www.mmpinc.org

Mountain States Meat Goat Association
Carrie Boyer
220 Chalk Creek
Coalville, UT 84017
www.msmga.org

Myotonic Goat Registry
PO Box 237
Chapel Hill, TN 37034
(931) 364-7206
www.myotonicgoatregistry.net
myotonic@myotonicgoatregistry.net

National Miniature Goat Association
6475 Fairfax Rd S
Bloomington, IN 47401
www.nmga.net
NminiaturegoatA@comcast.net

National MiniNubian Breeders Club
Kim Albano, Treasurer
156 Ironstone Dr S
Boyertown, PA 19512
www.mininubians.com
webmaster@MiniNubians.com

National Pygmy Goat Association
1932 –149th Ave SE
Snohomish, WA 98290
(425) 334-6506
www.npga-pygmy.com
npgaoffice@aol.com

National Saanen Breeders Association
PO Box 916
Santa Cruz, NM 87567
(505) 689-1371
www.nationalsaanenbreeders.com

National Toggenburg Club
Cathy Pindell
1156 East 4100 North
Buhl, ID 83316
www.nationaltoggclub.org
ntcinfo1@yahoo.com

New Mexico Meat Goat Association
PO Box 803
Moriarty, NM 87035
(505) 286-3631
www.nmmga.org
lkwilson@thuntek.net

New Mexico Pygmy Goat Club
www.geocities.com/nmpgc
nola@rusticrosefarm.com

New York State Dairy Goat Breeders Association
John Pfeiler, Treasurer
PO Box 181
Bloomfield, NY 14469
www.geocities.com/nysdgba

Nigerian Dwarf Goat Association
8370 Abbey Lane W
Wilhoit, AZ 86332
(928) 445-3423
www.ndga.org
ndgareg@aol.com

North American Packgoat Association
PO Box 170166
Boise, ID 83717
www.napga.org

North Arkansas Meat Goat Association
10591 Hwy 7 N
Harrison, AR 72601
(870) 429-5382
www.arkansasmeatgoat.com
namga@arkansasmeatgoat.com

North Carolina Meat Goat Association
Betty Herring
105 Five Bridge Rd
Clinton, NC 28328
www.ncmeatgoat.com

North Carolina Pygmy Goat Club
Dan Dawson
1205 Olive Branch Rd
Durham, NC 27703
www.angelfire.com/nc/ncpgc

North East Ohio Dairy Goat Association
www.neodga.com
neodga@neodga.com

Northeast Georgia Goat Producers Association
Gayle Shaw
1454 Bellamy Rd
Carnesville, GA 30521
www.nggpa.org
membership@nggpa.org

Northern Flinthills Dairy Goat Club
www.members.tripod.com/NFHDGC
drdeb@networksplus.net

Northern Kentucky Goat Producers
15650 Gardnerville Rd
Demossville, KY 41033
www.nkgp.com

Northern Lights Goat Association
Margie Vogel, Treasurer
W11047 Enterprise Lake Rd
Elcho, WI 54478
(715) 275-3532
www.geocities.com/northernlightsgoat

Northwest All Breed Goat Club
www.northwestgoatclub.com

Northwest Cashmere Association
Faith Hagenhofer
4342 Old Military Rd SE
Tenino, WA 98589
www.northwestcashmere.org

Northwest Oregon Dairy Goat Association
www.nwodga.org

Oberhasli Breeders of America
Elise Shope Anderson, Secretary
1035 Bardin Rd
Palatka, FL 32177
www.oberhasli.net
webmaster@oberhasli.net

Ohio Dairy Goat Association
Sue Davenport, Secretary
PO Box 199
Killbuck, OH 44673
www.odga.org

Ohio Meat Goat Association
Mary Morrow, Secretary
13140 Stoney Point Rd
New Concord, OH 43762
(740) 826-4333
www.ohiomeatgoatassociation.com
morrowfarm@aol.com

Ohio Valley Dairy Goat Association
www.kelpies.us/ovdga

Oklahoma Boer Goat Association
Dottie Wallace
Route 2, Box 14
Tryon, OK 74875
(918) 374-2445
www.obga.org

Oklahoma Meat Goat Association
41946 – 131st St W
Bristow, OK 74010
www.oklahomameatgoatassociation.com

Olympic Pygmy Goat Association
Jeff Smith
2206 186th Ave KPN
Lakebay, WA 98349
(253) 884-1331
www.opga-pygmy.com

Ontario Goat Breeders Association
RR #6
Dundalk, ON NOC 1BO
(519) 925-051
www.ogba.ca
lynnbandera@inetsonic.com

Pedigree International
3326 First Rd S
Humansville, MO 65674
(417) 754-1155
www.pedigreeinternational.com

Pennsylvania Club Livestock Association
Forrest Ohler, Treasurer
296 Dumbauld Rd
Rockwood, PA 15557
www.pacla.org

Pennsylvania Dairy Goat Association
Cathy Soult, President
1570 Keystone Way
Newport, PA 17074
(717) 567-9235
www.pdga.biz
president@pdga.biz

Purchase Area Goat Association
Jennie Lynn, Secretary
(270) 376-2976
www.geocities.com/purchaseareagoat/mayfield
purchaseareagoat@msn.com

Pygmy Goat Club (Great Britain)
www.pygmygoatclub.org

Pygora Breeders Association
538 Lamson Rd
Lysander, NY 13027
(315) 678-2812
www.pba-pygora.org
pbaregistrar@aol.com

Rare Breeds Canada
1-341 Clarkson Rd RR1
Castleton, ON K0K 1M0
(905) 344-7768
www.rarebreedscanada.ca
rbc@rarebreedscanada.ca

Rare Breeds International
Andreas Georgoudis
Aristotle University of Thessaloniki 54124
Thessaloniki, Greece
www.rarebreedsinternational.org
RBI_OFFICE@agro.auth.gr

Redwood Empire Dairy Goat Association
PO Box 1232
Rohnert Park, CA 94927
www.redga.org

Rhode Island Dairy Goat Association
Cecile Beauchemin, President
Moosup, CT 06354
www.geocities.com/ridga04
cbeauchemin@99main.com

Rocky Mountain Pygmy Goat Club
Evelyn Fox, Secretary
11665 Marble Front Rd
Caldwell, ID 83605
www.rmpgc.topcities.com
eff-3@hotmail.com

San Clemente Island Goat Association
3037 Halfway Rd
The Plains, VA 20198
(540) 687–8871
www.scigoats.org
postmaster@scigoats.org

Savanna Goat Information
Carol and Jerry Webb
14385 Rattlesnake Trail
Ivor, VA 23866
(757) 877-7407
www.ourfarmsite.com/web/goats/goatsavanna
agwebbcj@aol.com

Sierra Pacific Pygmy Goat Association
Barbara Williams, Membership
7424 Cedar Meadows Lane
Shingletown, CA 96088
(530) 474-5138
www.sppga.topcities.com

Silver State Pygmy Goat Association
Ray Hoyt, Treasurer
1355 Sanden Lane
Minden, NV 89423
www.sspga.org

Smoky Mountain Dairy Goat Association
Ken Everett, Membership
9303 Hill Rd
Knoxville, TN 37938
www.angelfire.com/tn2/smdga

South Central Texas Goat Club
Dorothy Weimann, Membership
27517 Rice Rd
Hockley, TX 77447
www.sctexgoatclub.org

South Western Ohio Dairy Goat Association
Ginger Jackson
6984 Thompson Rd
Cincinnati, OH 45247
(513) 385-7107
www.geocities.com/swodga2001

Southern Arizona Dairy Goat Association
Charlotte Gemiginiani, Secretary
11120 Calle Vaqueros E
Tucson, AZ 85749
www.sadga.org

Southern Pygmy Goat Club
Donna Conley, Secretary
11177 Bunker Hill Rd
Rockvale, TN 37153
www.softek.net/spgc

St. Lawrence Valley Dairy Goat Association
Winny Sachno, Membership
246 Hadley Rd
Potsdam, NY 13676
www.angelfire.com/ny4/slvdga

Sunbelt Goat Producers Cooperative, Inc.
PO Box 5726
Sandersville, GA 31082
(478) 553-1003
www.sunbeltgoat.com
info@sunbeltgoat.com

Tall Corn Meat Goat Wether Association
1959 Hwy 63
New Sharon, IA 50207
www.meatgoatwether.com

Texas Caprine Club
Lynn McAdoo, Treasurer
3636 County Rd 613
Alvarado, TX 76009
www.texascaprineclub.bizland.com

Texas Cashmere Association
Al Reed
3270 Neri Rd
Granbury, TX 76048
www.texascashmere.com

Texas Saanen Breeders Association
PO Box 14
Aquilla, TX 76622
www.texassaanenbreeders.org

Texas Sheep & Goat Raisers' Association
PO Box 2290
San Angelo, TX 76901
www.tsgra.com

U.S. Boer Goat Association
PO Box 663
Spicewood, TX 78669
(866) 668-7242
www.usbga.org

Virginia Angora Goat & Mohair Association
Charles Bodie, Secretary
1002 Still House Dr
Lexington, VA 24450
(540) 463-2808
www.angoragoats.com
bodie@rockbridge.net

Virginia Meat Goat Association
Mike Holland, Membership
26559 Ivey Tract Rd
Drewryville, VA 23844
(434) 348-7891
www.members.aol.com/vmgalink
HollandFrm@aol.com

Virginia State Dairy Goat Association
Jeanne Allen
7266 Covingtons Corner Rd
Bealeton, VA 22712
(540) 439-4246
www.vsdga.org
fallenfencefarm@yahoo.com

Washington County Meat Goat Association
PO Box 934
Sandersville, GA 31082
www.goat-a-rama.com

Willamette Pygmy Goat Club
Karen Vlask, Treasurer
2750 Mistletoe Rd
Dallas, OR 97338
(503) 623-2687
www.geocities.com/wpgclub
fspygmies@yahoo.com

Wisconsin Dairy Goat Association
Clara Hedrich, LaClare Farm
(920) 849-2926
www.wdga.org
laclare@tcei.com

EQUIPMENT AND
SUPPLY
DEALERS

These dealers have an Internet presence and/or a catalog commonly used by goat owners. Mail order and the Internet can offer greater convenience and possibly even lower prices than retail stores.

American Livestock Supply
613 Atlas Ave
Madison, WI 53714
(800) 356-0700
www.americanlivestock.com
alscattle@americanlivestock.com

Caprine Supply
PO Box Y
DeSoto, KS 66018
(913) 585-1191
Orders: (800) 646-7736
www.caprinesupply.com
info@caprinesupply.com

New England Cheesemaking Supply
PO Box 85
Ashfield, MA 01330
(413) 628-3808
www.cheesemaking.com
info@cheesemaking.com

Farmstead Health Supply
PO Box 985
Hillsborough, NC 27278
(919) 643-0300
www.farmsteadhealth.com
questions@farmsteadhealth.com

Hamby Dairy Supply
2402 Water St SW
Maysville, MO 64469
(800) 306-8937
www.hambydairysource.com
hds@ccp.com

Hoegger Supply Company
PO Box 331
Fayetteville, GA 30214
(800) 221-GOAT
www.hoeggergoatsupply.com

Jeffers Vet Supply
310 Saunders Rd W
Dothan, AL 36301
(800) 533-3377
www.jefferslivestock.com
customerservice@jefferspet.com

Khimaira Farms
2974 Stonyman Rd
Luray, VA 22835
(540) 743-4628
www.khimairafarm.com
Info@KhimairaFarm.com

KV Vet Supply Company
3190 N Rd
David City, NE 68632
(800) 423-8211
www.kvvet.com

Lion Edge Technologies (Farm Software)
7663 Duquesne Way S
Aurora, CO 80016
(720) 222-0681
www.lionedge.com

Nasco Farm & Ranch
(800) 558-9595
www.enasco.com/farmandranch

Northwest Pack Goats & Supplies
Rex & Terri Summerfield
2050 Wilson Creek Rd
Weippe, ID 83553
(888) PACKGOAT
www.northwestpackgoats.com
sales@northwestpackgoats.com

Omaha Vaccine Company
11143 Mockingbird Dr
Omaha, NE 68137
(800) 367-4444
www.omahavaccine.com
customerservice@petsuppliesdelivered.com

PBS Animal Health
2780 Richville Dr SE
Massillon, OH 44646
(800) 321-0235
www.pbsanimalhealth.com
info@pbsanimalhealth.com

Pipestone Veterinary Supply
PO Box 188
1300 Hwy 75 S
Pipestone, MN 56164
Orders: (800) 658-2523
Information: (507) 825-5687
www.pipevet.com
sheep@pipevet.com

Wild Wings Farms and Supply
(317) 873-3603
www.wildwingsfarmsandsupply.com
info@wildwingsfarmsandsupply.com

Shutterstock

• • • • • • • • • • • • • • • •

PRINT RESOURCES

The goat becomes the professor whenever teachers can't be found.

—Turkish saying

PERIODICALS

Goat Biz Magazine: PO Box 2694, San Angelo, TX 76902. www.goatbiz.com

Goat Rancher: 225 Hankins Rd, Sarah, MS 38665. www.goatrancher.com

Goat Tracks Magazine: 558 Park Ave, Logan, UT 84321. www.goattracksmagazine.com/index.html

Meat Goat Monthly News: Ranch Publishing, PO Box 2678, San Angelo, TX 76902. www.ranchmagazine.com/mgn.html

Ruminations: The Nigerian Dwarf and Mini-Goat Magazine: PO Box 859, Ashburnham, MA 01430. www.smallfarmgoat.com

The Dairy Goat Journal: 145 Industrial Dr, Medford, WI 54451. www.dairygoatjournal.com

The GOAT Magazine: 2268 County Rd 285, Gillett TX 78116. www.goatmagazine.com

United Caprine News: PO Box 328, Crowley, TX 76036. www.unitedcaprinenews.com

BOOKS

Belanger, Jerry. *Storey's Guide to Raising Dairy Goats.* North Adams, MA: Storey Publishing, 2001. The first book in my library, this book was originally titled *Raising Milk Goats the Modern Way.*

Boldrick, Lorrie. *Pygmy Goats: Management and Veterinary Care.* Orange, CA: All Pub Company, 1996. A great reference for those with small-breed goats.

Coleby, Pat. *Natural Goat Care.* Austin, TX: Acres USA, 2001. A look at alternative medicine for goats.

Dunn, Peter. *The Goatkeeper's Veterinary Book, 3rd ed.* Ipswich, Suffolk, UK: Old Pond Publishing, 2004.

Eddy, Carolyn. *Practical Goatpacking.* Eagle Creek, OR: Eagle Creek Packgoats, 1999.

Goatkeeping 101, 2nd ed. De Soto, KS: Caprine Supply, 1999.

Jaudas, Ulrich. *The New Goat Handbook.* Hauppauge, NY: Barron's Educational Series, 1989.

Levy, Juliette de Bairacle. *The Complete Herbal Handbook for Farm and Stable, 4th ed.* London: Faber & Faber, 1991.

MacKenzie, David and Ruth Goodwin. *Goat Husbandry, 5th ed.* London: Faber & Faber, 1993.

Mionczynski, John. *The Pack Goat.* Boulder, CO: Pruett Publishing, 1992. The pack-goat bible!

Mitcham, Stephanie and Allison. *The Angora Goat: Its History, Management and Diseases, 2nd ed.* Sumner, IA: Crane Creek Publications, 1999.

———. *The Meat Goat: Their History, Management and Diseases, 2nd ed.* Sumner, IA: Crane Creek Publications, 2006.

Pugh, D. G. *Sheep & Goat Medicine.* Philadelphia, PA: Saunders, 2001.

Smith, Mary and David Sherman. *Goat Medicine.* Hoboken, NJ: Wiley-Blackwell, 1994. Very technical and complete, and I use this book all the time.

INTERNET RESOURCES

G stands for goat and also for genius.
—Kenneth Rexroth

Agricultural Marketing Resource Center. www.agmrc.org. Provides links to state agricultural resource departments and marketing programs.

Amber Wave Goat Resources. www.amberwavespygmygoats.com.

Ancient Valley Ranch. www.ancientvalleyranch.com. Gives fun projects for driving and working goats.

AngoraGoat.com. www.angoragoat.com.

Bar None Meat Goats. www.barnonemeatgoats.com

Biology of the Goat. www.goatbiology.com. Includes parasite identification images.

Boer & Meat Goat Information Center www.boergoats.com.

Capricorn Cashmere. www.capcas.com.

Cybergoats. www.cybergoat.com. Lists goat classified ads, auctions, and veterinarians.

Fias Co Farm Enterprises. www.fiascofarm.com. Offers excellent resources from a holistic viewpoint.

Goat Connection. www.goatconnection.com. Sells goats, equipment, and gifts.

Goat Source. www.goatsource.com.

Goat Wisdom. www.goatwisdom.com.

Goat World. www.goatworld.com.

GoatFinder. www.goatfinder.com. Lists goat classified ads.

Goat-Link. www.goat-link.com.

Goats: Goat Health, Husbandry, and Management. http://lists.wsu.edu/mailman/listinfo/goats. This listserve out of Washington State University provides invaluable information on raising goats. Subscribe to be a member.

Goatweb. www.goatweb.com. Features a list of goat breeders, classified ads, a message board, and a show calendar.

Great Goats. www.greatgoats.com.

Irvine Mesa Charros 4-H Club. www.goats4h.com.

Jack & Anita Maudlin's Boer Goats. www.jackmauldin.com.

Kinne's Minis. www.kinne.net.

Langston University Research and Extension. www.luresext.edu.

Maryland Small Ruminant Page. www.sheepandgoat.com.

Merck Veterinary Manual. www.merckvetmanual.com.

Northwest Pack Goats and Supplies. www.northwestpackgoats.com.

Saanendoah. www.saanendoah.com. Offers insights into dairy goats.

Sheep and Goat Marketing. www.sheepgoatmarketing.info.

Tennessee Meat Goats. www.tennesseemeatgoats.com.

USDA Animal and Plant Health Inspection Service. www.aphis.usda.gov. Details U.S. state regulations on animal health and import between states and internationally.

· ·

GLOSSARY: HOW TO TALK GOAT

Goats can't talk. That's crazy!
—Brian Fellows, *Saturday Night Live*

Every field has its own special language and terminology. Capriculture (goat husbandry) is no exception. Because of the nature of living creatures, this glossary also contains some medical terminology that can help you translate what other goat owners or your veterinarian may be saying.

abomasum. In a ruminant's stomach, the fourth, or true, digestive compartment, which contains gastric juices and enzymes.

abortion. An abnormal or early termination of pregnancy.

abscess. An enclosed collection of pus found in tissues or organs. Usually a sign of infection, although some are sterile.

accredited herd. A herd of goats that has been annually tested for a specific disease (usually tuberculosis) and has been found free of that disease.

acidosis. An abnormal condition in which the rumen becomes too acidic, usually due to overconsumption of grain or rich forage or sudden dietary changes.

aflatoxin. A toxin produced by the molds *Aspergillus flavus* and *Aspergillus parasiticus* that can contaminate animal feed and cause serious problems.

afterbirth. The placenta and fetal membranes normally expelled from the doe's uterus within three to six hours of kidding.

anemia. A deficiency of red blood cells or hemoglobin in the blood, often a sign in goats of parasites and a need for deworming.

antibiotic. A medicine designed to kill or inhibit the growth of harmful microorganisms. It can be administered topically, orally, or by injection.

Billy goat. *Shutterstock*

antihelmenthic. A medicine used to remove or kill worms and other internal parasites.

antitoxin. An antitoxic serum used to prevent or treat diseases caused by biological toxins, such as tetanus.

artificial insemination (AI). A breeding technique that involves depositing buck semen into the doe without sexual contact. This technique allows for increased genetic diversity.

banding. Neutering a buckling by the use of an elastrator castration band at the base of the testicles.

billy. An informal term for an adult male goat, typically an older non-wether in a meat or fiber herd. Dairy producers consider it a negative term.

bleat. The cry of a goat.

bloat. An acute indigestion characterized by swelling of the stomach due to excess gas from overeating new feeds or fresh forage.

body condition scoring (BCS). An evaluation system used to estimate the physical condition of a goat, typically on a scale of 1 to 5 (thin to fat).

bolus. An antibiotic administered orally as a large pill or as an intravenous injection.

bots. A parasitic disease caused by infestation of the stomach or intestines with botfly larvae, which crawl in through the nasal passages.

brood doe. A doe bred for the purpose of continuing a desirable bloodline and genetics in her offspring.

browse. Tender young grasses, leaves, twigs, and other vegetation favored by goats.

brucellosis (*also* undulant fever, Malta fever, *or* **Mediterranean fever**). A disease caused by infection with bacteria of the *Brucella* group, frequently causing abortions in does and fever in humans.

buck. An adult male goat.

buckling. An immature male goat.

butterfat. The fat content, or cream, of milk, used to make butter.

butt. To strike with the head or horns. Butting is the preferred method among goats for settling dominance issues.

cabrito. Tender goat meat from a three-month-old milk-fed kid weighing 20 to 35 pounds, popular in Mexican dishes.

cape. Unshorn strip of hair left along a fiber goat's spine, about 6 to 8 inches wide, to protect the animal from becoming chilled after shearing.

capriculture. Goat husbandry, including all aspects of raising and breeding goats.

caprine. Of, relating to, or characteristic of a goat.

caprine arthritic encephalitis (CAE). A disease that causes inflammation of the joints and brain in goats, similar to AIDS in humans. CAE prevention techniques include hand-rearing to prevent transmission of the disease from dam to kid.

carrier. (*Genetics*) An animal that carries a recessive gene and can produce offspring with a genetic defect. (*Medicine*) An animal that shows no symptoms of a disease but harbors infectious organisms that can infect other animals.

caseous lymphadenitis (CL). A highly contagious disease characterized by abscesses on the lymph nodes.

cashmere. A luxury fabric made of the fine under wool of non-angora goats.

certified herd. A herd that has undergone an annual test for brucellosis and been found free of this disease.

chevon. Goat meat from a six- to nine-month-old kid weighing 50 to 60 pounds.

chivo. Goat meat from an older goat.

chlamydia. A microorganism that causes pneumonia, abortion, diarrhea, conjunctivitis, arthritis, and encephalitis in goats. There are many strains of this bacterium, which multiplies only in living cells and is spread by direct contact with fresh body secretions.

chlamydiosis. An infectious disease that causes abortion in does. Pregnant women should not handle aborted caprine materials.

clostridia. Anaerobic bacteria against which most goats are vaccinated. Commonly found in the environment, this genus of bacteria includes the organisms responsible for tetanus and enterotoxemia.

cocci. Bacteria that live in contaminated manure and cause coccidiosis in goats and other livestock. Some species (*Toxoplasma* and *Cryptosporidium*) are infectious to humans and other mammals; other types require a specific animal host and are not a problem for humans.

coccidiosis. A parasitic disease that destroys the lining of the small intestine, causing watery diarrhea, dehydration, and sometimes death. It is spread by ingestion of feces.

colostrum. The first milk produced by the doe after kidding; full of antibodies and minerals.

corpus luteum. A gland that develops in the ovary following ovulation. It secretes the hormone progesterone and helps regulate the doe's estrus cycle.

corticosteroid (*also* **corticoid**). A steroid—such as aldosterone, hydrocortisone, or cortisone—occurring in nature as a product of the adrenal cortex. Corticoids are given to goats as medication to relieve pain or inflammation.

crossbreeding. Mating a buck and doe of different breeds to produce a hybrid offspring.

cryptosporidiosis (*also* **crypto**). A parasitic organism that proliferates in the small intestine and causes diarrhea in goats and humans.

cud. Food that is subjected to bacterial action in the rumen and then regurgitated to the mouth for more chewing. Cud chewing is unique to ruminants.

cull. To remove from the herd goats that are unsound, physically inferior, or below average in production.

dam. A mother goat.

dental pad. The hard gum pad on the upper jaw that substitutes for top front teeth in goats.

disbud. To prevent the horns on a kid from growing by cauterizing the horn buds with a hot iron.

doe. An adult female goat.

doeling. A female goat under one year of age.

drench. To give a goat liquid medication by pouring into the mouth.

dry. No longer yielding milk.

drylot. A penned area for holding a herd for an extended period.

dual-purpose. A breed that serves two purposes, such as milk and meat.

elastrator. A tool used to apply heavy rubber bands to the scrotum of a kid for castration.

encephalitis. Inflammation of the brain, usually indicated in goats with severe signs such as fever, lack of coordination, and convulsions. Several diseases cause encephalitis, including polioencephalomalacia and listeriosis.

enteritis. Inflammation of the intestinal tract.

enterotoxemia. A systematic disease caused by a bacterial toxin, *Clostridium perfringens,* in the intestines. Misnamed "overeaters," enterotoxemia causes stomach cramps, diarrhea, and convulsions. It is often fatal without treatment. Kids in the first two weeks of life are subject to **enterotoxemia type C,** which causes bloody infection of the small intestine and rapid death; older goats are subject to **enterotoxemia type D,** which is triggered by rapid changes in diet, weather, or stress.

escutcheon. Arched area at the back of the udder below the perineum where the hair grows up and out instead of down.

estrus (*also* **heat**). The breeding period preceding ovulation and during which the doe is receptive to mating the buck, usually lasting 24 to 36 hours. The peak of estrus, when the doe is most receptive to breeding, is called "standing heat."

extended lactation (*also* **milking through**). The practice of milking a doe for more than one season without rebreeding.

fleece. The wool shorn from a fiber goat at one time.

flushing. Management practice of improving a doe's nutrition just before ovulation to increase the number of eggs produced.

forage. Food for goats that contains fiber, such as silage, hay, and pasture.

freshen. To begin to produce milk. A first-time dam is called a "first freshener."

gastroenteritis. Inflammation of the stomach and intestines.

grade. A goat produced by crossbreeding purebred stock with nonpedigreed stock. To upgrade or grade up is the sequential use of purebred animals over a series of generations to end with almost pure offspring.

helminth. A parasitic worm.

hybrid vigor. Increased vigor, meat production, longevity, reproduction, or other superior qualities arising from crossbreeding different goat breeds.

iodine. An antiseptic for wounds.

Johne's disease. A chronic wasting disease of ruminants that causes diarrhea.

ked. A bloodsucking tick that pierces the skin, causing damage to a goatskin pelt.

kemp. Short, hairy fibers found in mohair that have a hollow core and do not accept dye.

kid. A goat less than six months of age.

kidding. Bearing young: *The doe was kidding for the first time.*

lactation. The period during which a doe produces milk. In dairy goats, 305 days is the standard length of lactation.

landrace. A race of animals developed in the wild through natural selection and ideally suited to the environment in which they live.

liver fluke. A parasitic leaf-shaped worm that rolls up like a scroll and infests the bile ducts or liver.

lungworms. Parasitic roundworms that infest the respiratory tract and lung tissue.

mange. A chronic skin disease caused by mites that infest and damage the skin and hair.

manure. The dung of livestock, commonly used to fertilize soil. Also known as stool, droppings, waste, excrement, fecal matter, poop, and nanny berries.

mastitis. Inflammation of the udder caused by bacterial infection.

metritis. Inflammation of the uterus.

milker. A goat that produces milk, most often used in reference to a dairy doe.

milk fever (*also* **parturient paresis**). A disease that affects dairy goats just after giving birth and at the start of lactation. A substantial drop in the blood calcium level interferes with nerve transmission, causing partial or almost total paralysis.

milk replacer. An artificial milk substitute fed to kids as part of a CAE prevention program.

mohair. The fine hair of the Angora goat, which can be made into a luxury fabric; also, the fabric itself.

nanny. An informal term for a female goat, typically a dam. Dairy producers consider it a negative term.

nitrate poisoning. A condition caused by eating toxic levels of nitrates from plants containing an excess of this product.

nose bots. Tiny larvae that crawl into the nasal passages.

omasum. The third compartment of a ruminant's stomach.

open doe. A female goat that has not been bred or has not become pregnant after breeding.

overshot jaw (*also* **parrot mouth**). A congenital defect in which the upper jaw projects beyond the lower jaw.

paralumbar fossa. Soft, hollow area high on either side of the goat below the loin; literal translation is "the depression next to the lumbar vertebrae." Can be palpated to detect rumen activity.

parasite. An organism that lives on or in another organism. External parasites, such as fleas, infest the skin, hair, and nasal and ear passages. Internal parasites, such as worms, infest the stomach, lungs, and intestines.

parturition. The act of giving birth.

pinkeye. Inflammation of the eye, most often caused in goats by mycoplasma or chlamydia. Often highly contagious between goats and transmissible to humans.

polled. Naturally hornless. A polled goat has two bumps where horns would typically grow.

precocious milker. A doe that comes into milk without being bred.

pregnancy toxemia. A metabolic condition of pregnant does in which the blood contains toxins, generally caused by an energy-deficient diet during late pregnancy.

probiotic. A living organism, such as yeast, used to manipulate fermentation in the rumen.

purebred. A goat of unmixed lineage.

raddle marker. A color crayon in a harness or colored paste put on a buck while pasture breeding.

reticulum. The second compartment of a ruminant's stomach, lined with a membrane having honeycombed ridges that increase the surface for absorption.

roughage. Coarse, bulky plant matter that is high in fiber, such as hay and silage.

rumen. In a ruminant's stomach, the large first compartment, which contains microbes that break down forage and roughage.

ruminant. An animal that chews cud and has a four-compartment stomach. Goats, sheep, and cattle are ruminants.

scours. Diarrhea in livestock, usually severe and watery.

scurs. Incomplete horn growth resulting from inadequately removed horns.

shipping fever. A respiratory disease usually acquired by a goat during transport.

silage. Fodder prepared by storing and fermenting green forage plants in a silo or an airtight bag.

sire. A father goat.

sore mouth (*also* **contagious ecthyma** *or* **orf**). A highly contagious viral infection that causes scabs around the mouth, nostrils, and eyes and may affect the udders of lactating does.

stanchion (*also* **head gate**). A device used to secure goats in a stall or at a trough for feeding, milking, or work such as hoof trimming or artificial insemination.

strip. To remove the last drops of milk from the udder.

undershot jaw. A congenital defect in which the lower jaw projects beyond the upper jaw and the lower teeth extend past the upper dental pad.

urinary calculi. Stones of mineral salts in the urinary tract. Commonly found in wethers, urinary calculi are caused primarily by an imbalance of dietary calcium and phosphorus.

vaccination. An injection of a weakened or killed pathogen, used to stimulate antibody protection against the pathogen and produce immunity to a disease.

vaginal prolapse. Protrusion of the vagina in does in late pregnancy.

wattles. Hair-covered appendages of flesh hanging from the neck of some goats. Wattles serve no real function; they are thought to have formerly served as scent glands, but this has not been proven.

wether. A castrated male goat.

white muscle disease. A disease caused by a deficiency of selenium, Vitamin E, or both, characterized by the degeneration of skeletal and cardiac muscles.

yearling. A goat that is between six and fifteen months of age.

yolk. Wax or grease in a mohair fleece. Yolk protects the fiber from sun and chemical damage.

INDEX

● ● ● ● ● ● ● ● ● ● ● ● ● ● ● ● ● ● ●

4-H, 7, 14, 24, 150, 154
American Angora Goat Breeders Association (AAGBA), 17, 140
American Boer Goat Association, 18
American Dairy Goat Association (ADGA), 10, 12–16, 28
American Goat Society (AGS), 10, 14
American Kiko Goat Association, 19
American Livestock Breeds Conservancy (ALBC), 13, 15, 20, 22
American Veterinary Medical Association, 130
Arapawa Goat Breeders—USA, 22
Behavior, 30–33, 74–75
Biosecurity, 35, 39, 72, 92, 126
Blood testing, 92–93
Bolus, 62, 64, 81, 85, 112–113,
Breeding, 7–8, 14, 17–18, 20–21, 28, 34, 38, 40–42, 44, 46, 61–62,
    66–77, 87, 92, 95, 100, 147, 150, 153, 158, 160–161–163
        American goats, 10, 16
        artificial insemination, 70–71, 75–77
        Experimental goats, 10, 16
        Grade goats, 10
        Landrace breeds, 10, 20, 22–23
        methods, 68–71, 78–79
        Purebred goats, 10, 16–17, 66, 140
        visiting, leasing, and owning a buck, 72–73
Breeds,
    dairy goats,
        LaMancha, 8, 10, 12–13, 16, 36, 41–42, 48, 66, 86, 153
        Nigerian Dwarf, 10, 13–15, 18, 21–22, 40, 60, 66, 138, 144
        Nubian, 7–8, 13–15, 19–21, 40–41, 66, 71, 86, 90, 115,
            120, 145–146, 150, 155
        Swiss breeds, 8, 10, 14–16, 20, 22, 86
            Alpine, 10–13, 15
            Oberhasli, 10, 15, 22
            Saanen, 10, 13, 15–16, 19, 66
            Sable, 10, 16, 66
            Toggenburg, 10, 12–13, 16, 19, 157
    fiber goats,
        Angora, 8, 17–18, 31, 66, 75, 86, 138–143
        Cashmere, 8, 17–20, 138, 140–143
        Cashgora, 17–18, 138, 141
        Colored Angora, 18, 140
        Nigora, 18, 138, 140–141
        Pygora, 18, 138–141
    meat goats,
        Boer, 8, 18–20, 40, 66, 71, 95–96, 134
        Kiko, 19–20
        Myotonic (Tennessee fainting goat), 8, 19–20, 22, 71, 137
        Texas Wooden Leg, 19–20
        Savanna, 20
        Spanish meat goat, 20
    miniature goats,
        Pygmy, 8, 13, 18, 21, 66, 138
        Mini dairy goats, 21
        Kinder, 21
    rare breeds,
        Arapawa, 22
        Golden Guernsey, 22–23, 67
        San Clemente, 22–23
Buying,
    from breeders, 42
    from private parties, 42, 147
Caprine venipuncture, 93
Castration, 40, 71, 100–102, 110, 122, 137, 147
Choosing and buying, 40–45
Colored Angora Goat Breeders Association (CAGBA), 18, 140
Costs, 42–43
Culling, 27, 41–42, 66, 68, 87, 92, 105, 115, 136, 141, 158
Dairy goats, 6–10, 12–16, 20–23, 28–29, 34–35, 40, 55, 57, 63, 66,
    78, 87, 95, 114, 117, 136, 150–151, 159, 161
        commercial dairying, 6, 43, 63, 88, 128–131
        dairy barn, 43, 56, 130–131
        hand-milking, 132–133
        keeping, 128–133
        milk handling, 129–130
Dairy Herd Information Association (DHIA), 16
Dead animal disposal, 38–39
Elder care, 65
Electric fencing, 46, 48–49
Estrus, 71
Feeding, 35–37, 42, 44, 54–65, 67, 72, 88–89, 100, 103, 106,
    112–113, 124–125, 130–131, 135–136, 140, 142, 147, 157, 163
        grain, 36, 54, 56, 62–65, 78, 88, 119, 135, 147, 151
            feeders, 62–63, 151
        hay, 36, 54–56, 59, 61–62, 64, 78, 88–89, 120, 136, 151
            feeders, 52, 58–61, 151
        pasture or range, 34, 36, 40, 55–57, 62, 69, 89, 95, 135, 140
        silage, 58
        supplements, 34, 54, 56–57, 62–66, 80, 109, 123,
            135–136, 147
FFA, see National FFA Organization
Fiber goats, 8, 20, 34–35, 40, 42, 66, 69, 87, 95, 100, 103, 105, 136,
    151, 160–161
        harvesting, 140
        keeping, 138–143
        markets, 143
        mohair, 17–18, 66, 138–139, 141–143
Flies, 53, 96, 131, 142
Food and Agriculture Organization (FAO), 21, 124
Fund for Animals, 23
GOATS list, 41
Hides and leather, 8–9, 103, 135, 158, 160
Hoof trimming, 92–94, 110, 120, 151, 163
Horns, 9, 58, 60, 72, 74–75, 92, 94–98, 124, 144, 148
        disbudding, 43, 77, 92, 94–99, 110, 122, 147
Housing, 34, 36, 40, 44, 46–53, 72–73, 87, 92, 124, 163
        shelter, 50–52
Humane Farm Animal Care (HFAC), 34

Identification, 35–36, 87, 126, 140, 162
    ear tag, 35–36, 72, 87, 93, 126, 140, 162
    tattooing, 35–37, 72, 87, 92–93, 140, 153, 162
Injections, 64, 87, 99–100, 110–112, 114, 117, 119–121
International Dairy Goat Record Association, 10
International export, 126
Interstate shipping, 125
Kidding, 38, 44, 52, 54, 58, 62, 66–67, 71, 77–91, 95, 104, 117, 157
    abnormal situations, 84
    bottle-rearing, 43, 77, 86–89, 115, 117, 137
    care during pregnancy, 62, 64, 67, 72, 78–80, 124
    colostrum, 78–79, 81, 83, 88–91, 105, 112, 117, 119
    dam-raising, 77, 83, 87–88
    injections, 87
    labor, 79
    pan feeding, 90
    supplies, 80–81
Livestock guardians (LG), 98–100
Management systems, 20, 36–38, 42, 92–93, 98, 104–105, 115, 117, 140, 157, 163
    feedlot management, 35, 37, 134–135
    organic, 36–38, 105–106, 163
    sustainable agriculture, 37–38, 105
Marketing, 42, 66, 77, 87, 100, 134–136, 143, 158–163
    association links, 160
    auctions, 35, 41–42, 44, 136, 147, 158
    direct, 136, 158–160
    ethnic markets, 8, 134, 137, 158
    sales ethics, 162–163
Mating systems, 68
Meat goats, 7–9, 13–14, 17, 19–20, 34–35, 40, 42, 66, 69, 87, 95, 100, 104, 136, 151
    carcass characteristics, 134–135
    feeding, 135–136
    keeping, 134–137
    markets, 136–137, 143
Musical instruments, 9
National Agricultural Statistics Services All-Goat Survey 2007, 17
National Animal Identification System (NAIS), 35, 126
National FFA Organization, 7, 24, 150
Neutering, 43, 77, 92, 100–103, 147
Pack and harness goats, 8, 95, 148–149
Pasteurizing, 90–91, 115, 117, 130, 151
Pet goats, 8, 13–14, 20, 34, 40, 43–44, 56, 58, 62, 66, 87, 95, 100–101, 138, 144–149, 159–160, 163
Physical characteristics, 10–29
    Body condition scoring, 28–29
    Coat patterns, 11, 16, 20–21
    Digestive system, 28, 107
    Mammary system, 25
    Mouth, 27, 65
Pygora Breeders Association, 18, 139
Record keeping, 35–36, 126, 151, 163
Registering, 43, 45, 72–73, 93, 140, 150–151, 154, 158, 162–163
Regulations, 34–37, 158, 162

Ruminants, 28, 30, 50, 54–55, 64, 104, 107, 109, 118–119, 124, 135–136
Sheep, 30–33, 46, 62, 115–116, 120, 140, 145
Showing, 7, 35, 43, 45, 54, 66, 72, 92, 115, 117, 122, 124, 131, 136, 140, 150–157, 160–161, 163
    clipping, 150–153, 155
    finding shows, 150–151
    judging, 154, 156–157, 160
    tips, 157
Sickness, 35, 42, 44, 50, 52, 54, 56, 58, 62–64, 72–73, 84, 87–88, 91, 105–124, 153, 163
    CAE (Caprine arthritis encephalitis), 43, 83, 86, 88, 90, 92, 114–117, 120, 163
    chronic, 92, 115
    CL (caseous lymphadenitis), 87, 110–111, 115, 117–118
    common diseases, 115–123
    diarrhea, 44, 54, 62, 72, 109, 118–120, 153
    entropion, 119–120
    finding a veterinarian, 114–115
    floppy kid syndrome, 120
    foot rot, 120
    frothy bloat, 120
    G6S deficiency (glucosamine-6-sulfatase), 115, 120
    hoof-and-mouth disease, 115, 120
    humane slaughter, 114
    Johnes disease, 88, 115, 120
    mastitis, 110, 121, 130–133
    medicine, 36, 63, 66, 78, 80–81, 85, 100, 104, 106, 109–113, 115–116, 120–123
    neurological problems, 121
    observation and examination, 106–109
    parasite prevention, 20, 36–37, 50, 54, 63, 85, 88, 92, 103–104, 109–110, 115, 118, 136, 142, 157, 163
    pinkeye (infectious keratoconjunctivitis), 121
    respiratory illness and pneumonia, 31, 44, 62, 72, 113, 117, 121–122, 153
    ringworm, 72, 103, 110, 122, 153
    scrapie, 39, 115, 122
    sore mouth, 105, 122–123
    stomach tubes, 112–113
    tetanus, 87, 96–97, 99, 101–102, 105, 122–123
    tuberculosis, 35, 115, 120, 123, 130
    urinary calculi, 62, 64, 100, 123, 136, 147
Terrapin Acres, 6–7
Traveling with goats, 92, 103, 124–127, 151
U.S. Animal Health Association, 116
United States Department of Agriculture (USDA), 8, 18, 35, 125
Vaccination, 42, 77, 79, 87, 92, 97, 102, 104–105, 117–118, 122–123, 147, 157, 163
Veterinarian, 35, 42, 71–73, 78, 80–81, 84, 92, 95–96, 100–101, 103–105, 109–110, 112, 114–116, 118, 120–125, 128, 136
Watering, 34, 54–55, 65, 89, 104, 106, 112–113, 123–125, 130, 142
Wether, 29, 40, 64, 71–72, 77, 100, 103, 123, 135–137, 147, 161

ABOUT
THE
AUTHOR

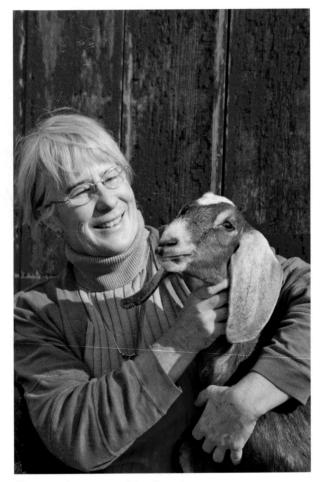

Photograph courtesy of Jen Brown

Author Carol A. Amundson has been raising goats since 1989. Her articles have appeared in *Goat Magazine* and *United Caprine News*, and she is the former editor of the Minnesota Dairy Goat Association newsletter, *Gopher Goat Gossip*. Carol lives with her husband, Wayne, and daughter, Viveka, on Terrapin Acres, a twenty-acre farm outside of Scandia, Minnesota.